Taphonomy and Interpretation

Symposia of the Association for Environmental Archaeology No. 14

Edited by Jacqueli

Published by
Oxbow Books, Park End Place, Oxford OX1 1HN

ISBN 1 84217 004 X

This book is available direct from
Oxbow Books, Park End Place, Oxford OX1 1HN
(Phone: 01865–241249; Fax: 01865–794449)

and

The David Brown Book Company
PO Box 511, Oakville, CT 06779, USA
(Phone: 860–945–9329; Fax: 860–945–9468)

and

via our website
www.oxbowbooks.com

Printed in Great Britain at
The Short Run Press
Exeter

Contents

AEA 1993 conference at Durham: editorial

Sue Stallibrass

These papers were originally presented at the 1993 annual conference of the Association for Environmental Archaeology, held at Durham University 18th–21st September 1993. Not all of the presented papers and posters are represented in this volume, but the full list is given at the end of the volume.

The themes of the conference were taphonomy and interpretation, to encourage speakers to go beyond data acquisition and description and to question:

a) how material (pollen, insects, bones etc.) came to be deposited in the contexts from which they were recovered,
b) how surviving material might compare with what existed in the past,
c) how our own methodologies can bias our results, and
d) how material might be interpreted in terms of past human activities and environmental processes.

Not so much just "What is it?" as "What was it and what does it all mean?".

The themes are relevant to all archaeological and palaeoecological enquiries, regardless of the type of material studied, the original date of deposition or the nature of the site. There is a good spread of materials considered: at least two papers each on pollen, invertebrates, carbonised plant remains, sediments and animal bones, and the taphonomic processes reviewed include people, coffin flies, conflagrations and statistics. The archaeological periods studied range from the Bronze Age to the 19th century AD, and include rural, urban and burial sites.

The approaches utilised range from controlled experiments, through observations of modern and recent analogue situations, to detailed considerations of the archaeological data and the methods used in their investigations. These methods, in turn, range from the broader issues of combining, in an interactive way, different lines of evidence, through to detailed examination of methods of recording, identification and excavation.

Despite the wide range of topics and approaches, the papers contain recurrent themes. Several authors advise against the presumption that data provide unbiased samples of past populations, and demonstrate some of the tests that can be applied. Even biased samples can be interpreted with some confidence, provided that those biases can be identified, and have been demonstrated to act in predictable, systematic ways. The advice is: exercise caution, and be realistic. Even if a sample is biased to such an extent, or in such a way as to make it unsuitable for interpretation in terms of its original nature or mode of deposition (i.e. what it all meant), the identification of taphonomic changes can lead to an understanding of the dynamic processes that have been active at the site (i.e. what's happened to it since). This can lead to greater selectivity in terms of sites and/or material suitable for investigating specific research questions, and to a more informed approach to the management of archaeological sites.

Many authors are concerned with the identification of patterns in the data. Some work with modern material, looking to see if certain processes lead to distinctive 'fingerprints': diagnostic suites of characteristics. As often as not, such 'fingerprints' are not diagnostic at all, but can be created by a range of different processes whose effects are indistinguishable from each other. This does not mean that interpretation is not feasible, simply that a range of options must be considered. Others work with archaeological data, attempting to interpret what they see, not to predict what they might find. Rigorous methodologies are recommended, and there are several warnings that techniques can influence results. Others try to distinguish between the absence of evidence and evidence of absence, and demonstrate that patterns can be more apparent than real.

Three of the actualistic studies warn against the perils of over-simplistic one to one interpretations. Stokes' investigation of a distinctive pattern of butchery of cattle limb bones demonstrates that a single pattern of butchery can be associated with several different processes, each leading to different products with different uses. Nor is the epigraphic evidence any less ambiguous: relevant words in classical sources have been translated with a variety of different meanings. The lesson here is to keep an open mind when interpreting archaeological materials, and to be aware of ranges of possibilities.

Dobney, Hall and Hill's observations of modern chicken behaviour also remind us that the effects of several different processes can converge, in this case to produce similar, rather nebulous features: is it a pit, a tree hole or a chicken bath? In a second example, they note that an association of certain plant and invertebrate species need not be the remains of stable manure, but can simply reflect someone's attempt to avoid getting their feet wet!

David Smith's study of modern coleoptera in a traditional farm also demonstrates that several different processes or activities can lead to similar associations, in this case faunal suites. In addition, some habitats are 'invisible' with very few coleopteran inhabitants, whilst others suites of fauna, reflecting specific habitats, fail to be overprinted by changing circumstances. A sample from one habitat contained two suites of fauna, suggesting that more than one microhabitat had (unintentionally) been sampled. If we cannot discriminate between different habitats whilst they are still in existence, how can we attempt to identify them in the past? Nonetheless, Smith's conclusion is optimistic. Rather than throw the baby out with the bathwater, he emphasises the importance of considering data at a suitable scale (in this case spatial, though a temporal scale can be equally important). He finds that a detailed, specific question such as 'Was this building used to store hay?' is not easily answerable, even in a modern situation, but that a more general question such as 'Was there hay at this site?' can be addressed. He also comments that, whilst an appreciation of taphonomic complexities should put us on guard against over-simplistic one to one equations of data and interpretations, at the same time it can enhance our understanding of processes that have been active at a site in the past.

Although convergent processes may sometimes be distinguishable by the application of additional lines of enquiry, in many archaeological cases we must be prepared to keep an open mind and acknowledge the fact that we cannot say with certainty which of several 'just so' stories is the most realistic interpretation. The variety of possible interpretations is one of the aspects of archaeology that makes it so much fun: just like a good poem that can be read in many ways. What we have to be wary of, however, is that we study the original, and do not make it up for ourselves.

Three other actualistic studies suggest that, in some cases, the correlation of cause and effect can be relatively restricted. Hakbijl looks at arthropods found in association with 19th century human burials. Because the manner of the burials is known, and the arthropods have known life cycles and obligatory or facultative habitats, certain associations can be used as indicators of burial practices, even when other evidence does not survive. They might, for instance, indicate whether or not a corpse was partially decomposed prior to burial, or whether a wooden coffin was used.

In a similar manner, Wijgaarden-Bakker's long term study of a buried ram corpse is producing information regarding the rates and sequence of carcase decomposition and bone diagenesis in a monitored burial environment. These studies give us useful guidelines for interpretation of similar evidence surviving from the past.

Nicholson's study of fish bones deriving from modern otter spraints deliberately considers alternative formation processes, in order to test for convergent assemblage types. Having demonstrated that one taphonomic agent can produce faunal assemblages that share many common characteristics with the selected prehistoric samples, she then checks alternative agents (seals and seawater) but shows that they result in assemblages with rather different characteristics, thus confirming otters as the (currently) favoured agent of deposition for the archaeological material.

Several contributors recommend the use of multiple lines of enquiry as a means of counteracting the biases inherent in any one type of approach or group of material. Carter and Holden put this recommendation into practice with a combined macrobotanical and sedimentological consideration of prehistoric crop husbandry. Importantly, they address explicit questions, clearly separating the evidence from their interpretations. They consider taphonomic biases that might have led to apparent, rather than real, differences in the evidence pertaining to two different periods of occupation, and suggest further lines of enquiry that might resolve continuing uncertainties.

Rowley-Conwy's consideration of prehistoric macrobotanical remains is also primarily interested in ascertaining whether differences are real or apparent. Rather than comparing two periods at a single site, he restricts his considerations to the Iron Age, but compares data from different geographical regions: particularly Britain and Denmark. He concludes that some differences are 'non-real' in archaeological terms but have been caused by differing methods of excavation, recovery and analysis. However, other differences cannot be explained in this way, and do appear to represent different taphonomic pathways that were active at the time that the sites were occupied. He relates these differences to the structure types in which the remains were found which, in turn, had implications for prehistoric methods of storage, and potential processes of destruction and preservation. All things being equal, grain in an Iron Age house in Britain should have had the same chances of being preserved for recovery through excavation as grain in an Iron Age house in continental Europe. But all things were not equal, and Rowley-Conwy's paper illustrates the need to use several lines of enquiry, not all of them necessarily considered to be 'environmental' or 'palaeoeconomical' in approach.

The potential for preservation is an important aspect of pollen studies, and Melanie Smith's paper is a detailed consideration of how to define a sampling strategy to address specific questions relating to past landscapes. Building upon what is already known about taphonomic biases in pollen recruitment (catchment areas, modes of transport, rates of production etc.), she demonstrates how to select sampling sites that are most likely to produce data that have predictable, rather than unknown, taphonomic biases. Since reality is not designed to fit statistical sampling strategies, she also discusses the decisions that had to be made in order to achieve a 'best fit' between potential and ideal.

Several papers are concerned with the nature of the recovered archaeological data, looking for patterns that might be interpretable either in terms of past environments (or economic activities), or in terms of taphonomic processes.

Tipping's study is concerned with what might happen to pollen once it has been deposited. He observes that pollen from archaeological sites is seldom recovered from deposits with ideal preservation conditions, and calls for strong caution before a death assemblage (thanatocoenosis) is assumed to be representative of a living community (biocoenosis). He urges the use of routine recording of states of preservation to test whether or not a palaeoecological interpretation can be justified. He also illustrates how intellectual approaches can direct, restrict or bias interpretations. States of preservation can be considered to represent a hierarchy (from poor to excellent): this facilitates quantitative and statistical analysis, but may be based upon a false premise. Alternatively, states of preservation can be considered to reflect the consequences of different taphonomic processes, in which case each state has to be considered in its own right. Until more actualistic work is undertaken to study which processes can lead to which (if any) diagnostic preservation states, the choice of analytical methodology remains rather a matter of personal philosophical choice.

Sidell's study of a dark earth investigates the degree and nature of taphonomic biases caused by the reworking of deposits by pedogenetic processes. The methodology is similar to that used in studies of distributions of materials in plough-soils, and it produces some encouraging results. Dark earths often obscure the final phases of activity at a site before a period of abandonment or change of use. Sidell's study demonstrates that, although a deposit may initially appear featureless and relatively un-interpretable to the excavator, controlled excavation and detailed analysis can reveal useful evidence that is systematically altered, rather than obliterated.

Patterns of data are also searched for by Moreno Garcia and Rackham in their post-excavation analysis of animal bones. Using a non-predictive statistical model originally designed for use with pottery sherds they look for associations between species and anatomical elements. They then consider likely causes for these patterns, which range from methodological biases at the level of excavation and analysis to original biases in butchery methods and locations of disposal. Although the method can point out the obvious (e.g. the reluctance of specialists to identify vertebrae to species level) it can also highlight idiosyncrasies that might be hidden at a broader level of analysis (such as the concentrations of horse bones on sites that were marginal to Roman urban centres). Their paper serves to remind us that our own perspectives and methodologies can bias interpretations just as much as past human activities or site formation processes.

Wilson's paper considered the most complex and effective taphonomic agent of them all: human nature. He doesn't attempt to address how one might relate observed characteristics or patterns in archaeological data to people's thoughts, attitudes and approaches to life, but highlights the fact that not all attributes of archaeological data are linked to pragmatic, utilitarian activities. In particular, he looks at how a society might behave rather than an individual.

The papers address taphonomy at all stages: from site formation processes through diagenesis and sampling procedures to analytical techniques and interpretative perspectives. Curiously, all six of the studies that investigate possible analogue data concern faunal remains. Perhaps this is simply a sampling error, pointing out yet another recurrent theme of this conference: the question of scale. Specific questions need to be addressed to appropriate sample sizes and at an appropriate level of detail. The data, the methods of recovery and the analytical techniques all have to be considered for their quality and relevance to the questions being asked. Since the data are what they are (not what we might like them to be), if they do not comply with our prerequisites then we have two choices: we can either reject the data or change our questions.

The authors of all of these papers demonstrate that taphonomic biases should not be regarded as insuperable and obligatory destroyers of archaeological data. In many cases, they should be perceived as warping or biasing data in systematic ways. Provided that these systematic biases can be identified (and, if possible, quantified), then they can be taken into consideration before any interpretations are made.

To be unable to say, with certainty, exactly what happened at a particular point in time is not, necessarily, an admission of defeat, but a recognition of the diversity of the past and of the intervention of other processes and events between that point and now. In many cases, those intervening processes and events are of great interest in themselves: like all landscapes, archaeological sites are dynamic

The overall tenet is caution and prudence rather than despair, together with an acknowledgement that archaeological data are complex for a variety of reasons, some relating to the original data, some to intermediary transformation processes, and some caused by our own perspectives, biases and technology.

Editors' note: Two of the papers in this volume discuss work undertaken as part of a (then) ongoing project at Lairg, Scotland. Since the conference, a major monograph has been published describing the Lairg project: McCullagh, R.P.J. & Tipping, R. 1998. The Lairg project 1988–1996. *The evolution of an archaeological landscape in northern Scotland*. Leith: *Scottish Trust for Archaeological Research* Monograph 3.

List of Contributors

STEPHEN P. CARTER
Headland Archaeology Ltd
Unit B4
Albion Business Centre
78 Albion Road
Edinburgh EH7 5QZ

KEITH DOBNEY
Environmental Archaeology Unit
Department of Biology
PO Box 373
York YO10 5YW

TOM HAKBIJL
Zoological Museum
Department of Entomology
Plantage Middenlaa 64
1018 DH Amsterdam
The Netherlands

ALLAN HALL
Environmental Archaeology Unit
Department of Biology
PO Box 373
York YO10 5YW

MICHAEL HILL
Environmental Archaeology Unit
Department of Biology
PO Box 373
York YO10 5YW

TIMOTHY G. HOLDEN
Headland Archaeology Ltd
Unit B4
Albion Business Centre
78 Albion Road
Edinburgh EH7 5QZ

MARTA MORENO GARCIA
Faunal Remains Unit
Department of Archaeology
University of Cambridge
Cambridge CB2 3DZ

REBECCA A. NICHOLSON
Department of Archaeological Sciences
University of Bradford
Bradford BD7 1DP

JAMES RACKHAM
25 Main Street
South Rauceby
Lincolnshire NG34 8QG

PETER ROWLEY-CONWY
Department of Archaeology
University of Durham
Science Laboratories
South Road
Durham DH1 3LE

ELIZABETH J. SIDELL
Institute of Archaeology
University College
31–34 Gordon Square
London WC1H 0PY

DAVID N. SMITH
Department of Ancient History and Archaeology
University of Birmingham
Edgbaston
Birmingham B15 2TT

MELANIE SMITH
School of Agriculture
De Monfort University
Riseholme Hall
Lincoln LN2 2LG

P. R. G. STOKES
Department of Archaeology
Universtiy of Durham
Science Laboratories
South Road
Durham DH1 3LE

RICHARD TIPPING
Department of Environmental Science
University of Stirling
Stirling FK9 4LA

LOUISE VAN WIJNGAARDEN-BAKKER
Institute for Pre-and Protohistory
University of Amsterdam
Nieuwe Prinsengracht 130
1018 VZ Amsterdam
The Netherlands

BOB WILSON
8 The Holt
Abingdon
Oxfordshire OX14 2DR

1. Interpreting prehistoric cultivation using the combined evidence of plant remains and soils: an example from northern Scotland

Stephen P. Carter and Timothy G. Holden

SUMMARY

Information relevant to the study of most topics in archaeology can be obtained from more than one source and, in general, a multidisciplinary approach to a problem will yield a better result. This is particularly true when problems of taphonomy hinder interpretation of individual sources of data. In this paper, analyses of soils and charred plant remains are used in an attempt to illuminate aspects of prehistoric cultivation at Lairg in northern Scotland. The results have highlighted significant differences in arable cultivation practice between the Bronze Age and Iron Age settlements excavated so far. Nevertheless, if these are put into a broader archaeological framework, interpretation becomes more problematic. At this level several equally valid explanations can be constructed. These are outlined and further lines of investigation which may help to separate them are suggested.

1. INTRODUCTION

Many of the present data relating to prehistoric agricultural systems derive from field observation and archaeological excavation. Features such as banks and ditches, once used to define the boundaries of fields, are often encountered. Accumulations of stone and soil formed by repeated cultivation of the same area form distinctive landmarks such as lynchets and clearance cairns. More elusive features such as cultivation ridges, ard and spade marks also survive occasionally to reveal evidence of particular cultivation techniques.

However, experience in Scotland has revealed a very complicated situation in which changing land use and agricultural features leave a bewildering palimpsest of archaeological remains. Earlier structures are destroyed during the construction of later ones and frequently obliterated by later agricultural practices. Older features are re-used and incorporated into new designs. Many cannot accurately be dated and cannot therefore be linked to the settlements which they served. Features such as ard-marks and cultivation ridges rarely survive unless they are succeeded by total abandonment of the fields or protected by substantial structures such as houses or burial cairns built upon their surfaces. Surviving examples are frequently located only in areas which were abandoned and have remained uncultivated in all subsequent periods. There are numerous possible reasons for such abandonment but in many cases it was probably because they were marginal with respect to the cultivation of crops. In such cases the surviving evidence is of fields which failed and are, in this sense, atypical of the larger scheme of things. In summary, the surviving archaeological evidence tends to be diffuse, difficult to date or link stratigraphically and riddled with taphonomic problems.

Archaeological science can add significantly to our understanding of these fields through the study of plant remains and cultivated soils. Each of these is, however, subject to its own suite of taphonomic problems. Plant remains are biased in favour of those crops which require the use of heat as part of their processing cycle. Whole categories of potential economic plants, such as vegetable crops, do not survive well if charred and many are not readily identifiable even if they are charred by contact with fire. Soils tend to represent only the latest phase of activity on the site and frequently incorporate reworked earlier soils. Following abandonment, pedogenesis continues to modify the agricultural soils such that even where soil surfaces are protected their structure today is very different to that created in antiquity.

This possibly paints an unnecessarily grim picture but amply highlights the complexity of archaeological taphonomy with respect to prehistoric agriculture. It is, however, apparent that even though each line of investigation has its limitations, the data are nevertheless complementary. It is therefore the intention of this paper to draw together

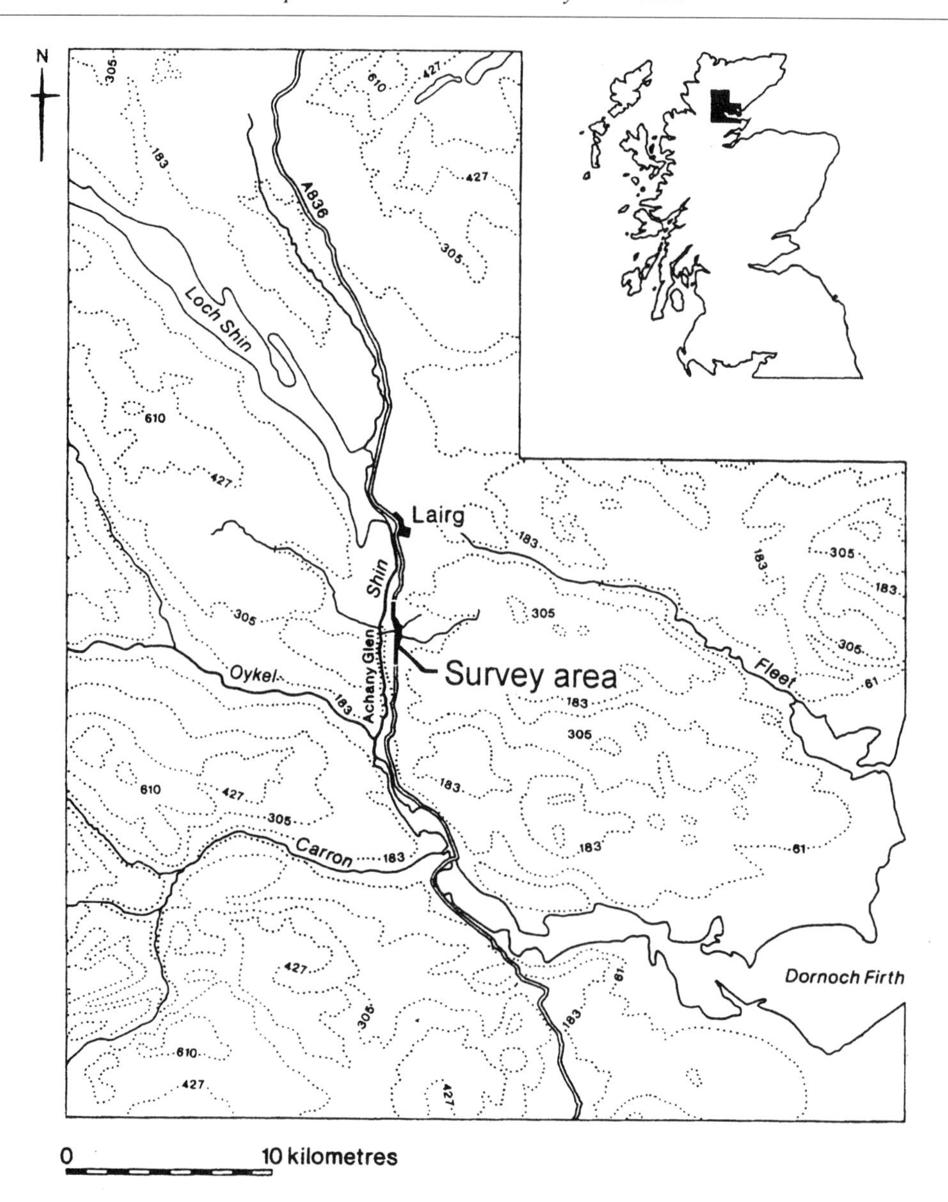

Figure 1.1: Location of the survey area.

some of these disparate and imperfect data in an attempt to illuminate aspects of prehistoric agriculture with specific reference to archaeological excavations near Lairg in northern Scotland.

2. THE EXCAVATIONS AT LAIRG

An initial survey of a 3 km road corridor identified over 600 field monuments reflecting widespread settlement and arable agriculture (McCullagh 1992, 1993 – Figures 1.1 & 1.2). Excavations examined a total of twenty-two of these field monuments including round-houses, burnt mounds, clearance cairns, field-edge banks and substantial areas of cultivated fields (McCullagh and Tipping 1998). They span the period from 2500 Cal BC to the recent past. The pattern of dates and the archaeological evidence indicate that a Neolithic landscape, from which only a few massive chambered tombs survive as visible monuments, was subsumed into, and largely erased by, a vigorous and apparently expanding Bronze Age agricultural regime. What happened after 1000 BC is uncertain but its effect can be seen in the reduction of recognisable settlement sites and later on by the appearance of defended sites.

The results of a substantial radiocarbon dating programme indicated that the cultivated soils and plant remains could be assigned to two time periods. These were parts of the Bronze Age (*ca.* 1800 to 900 Cal BC) and the Iron Age (*ca.* 200 Cal BC to 500 Cal AD). It was also clear at an early stage, from field evidence, that the techniques of cultivation used in these two periods were apparently different. Data from the study of the soils and carbonised plant remains are likely to provide some of the most relevant information regarding the cultivation prac-

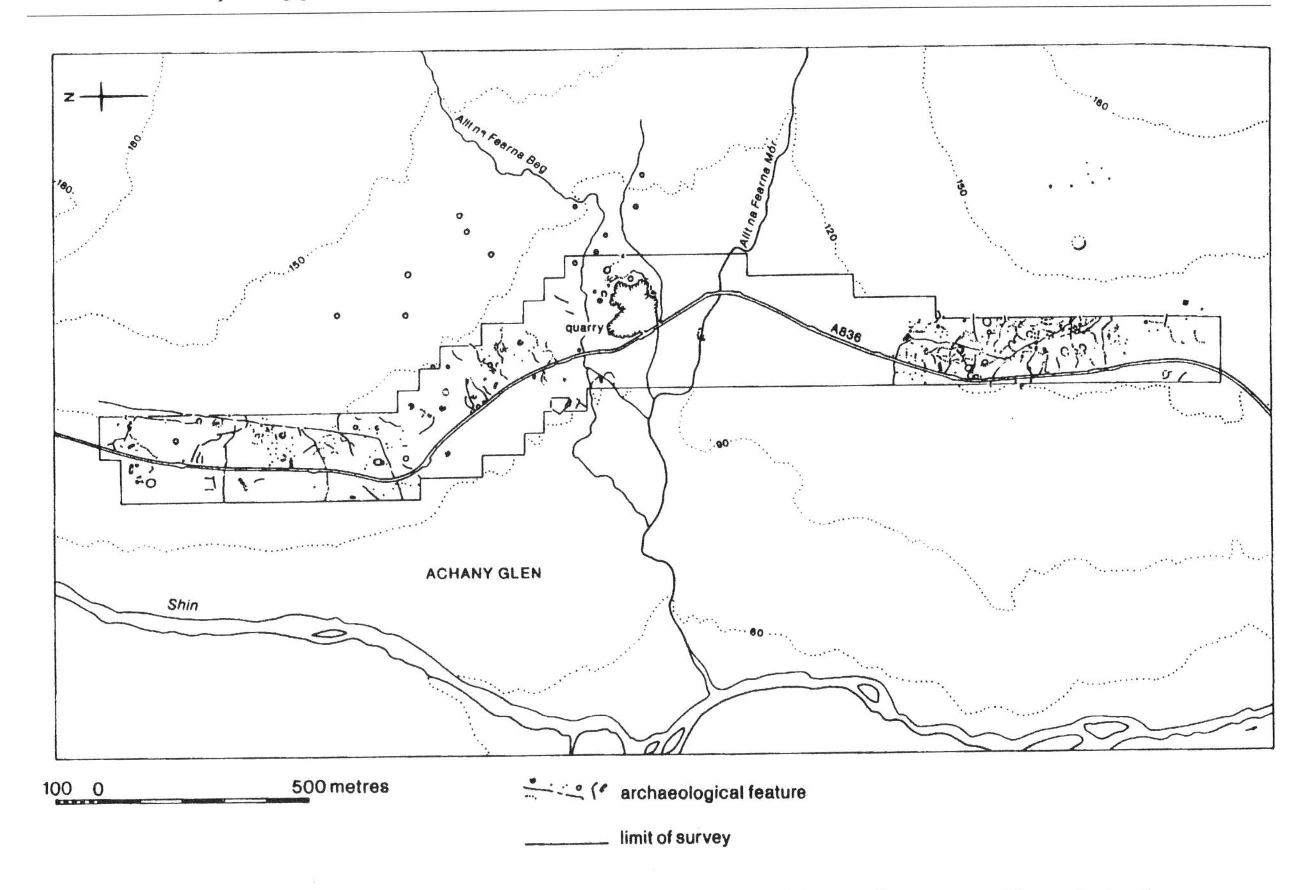

Figure 1.2: Distribution of archaeological features in the survey area. The results presented here derive from excavations to the east of the quarry.

tices used at Lairg so the following research questions were addressed.

1. Do the data from the analysis of the soil and charred plant remains indicate that different methods of cultivation were being used in the Bronze Age and Iron Age?
2. Are these differences real or a product of differential preservation and biased selection of sites for excavation?
3. If the apparent differences in cultivation technique are real then why did the method of cultivation change and what social and environmental processes brought about this change?

3. METHODS

Samples were collected from all sediment contexts for the recovery of plant remains by flotation and routine chemical and physical soil analyses. Thin sections were prepared from selected buried soils and other sediments where appropriate.

4. BRONZE AGE ARABLE AGRICULTURE AT LAIRG

(a) Evidence from the plant remains

A number of structures from the Bronze Age produced identifiable plant remains; these include a burial cairn, two burnt mounds and six round-houses. Samples from the round-houses present a series of domestic contexts comprising dumps, post-holes and floor surfaces and include at least two conflagration events when round-houses were burnt down (Houses 1 and 4 – see Figure 1.3). The samples from the cairn and burnt mounds add to the broader picture and act as controls for the domestic contexts.

Crop remains are very much dominated by naked barley with occasional examples of hulled barley (see Table 1.1). The cereals are, with the exception of only two barley rachis internodes identified during the whole project, represented by the grains alone. The concentration of these is consistent with small quantities of relatively clean grain dropped or accidentally charred during food preparation or storage. From the conflagration deposits, a much larger quantity of grain survives and is interpreted as grain that was being stored in the round-house, becoming charred when it burnt down.

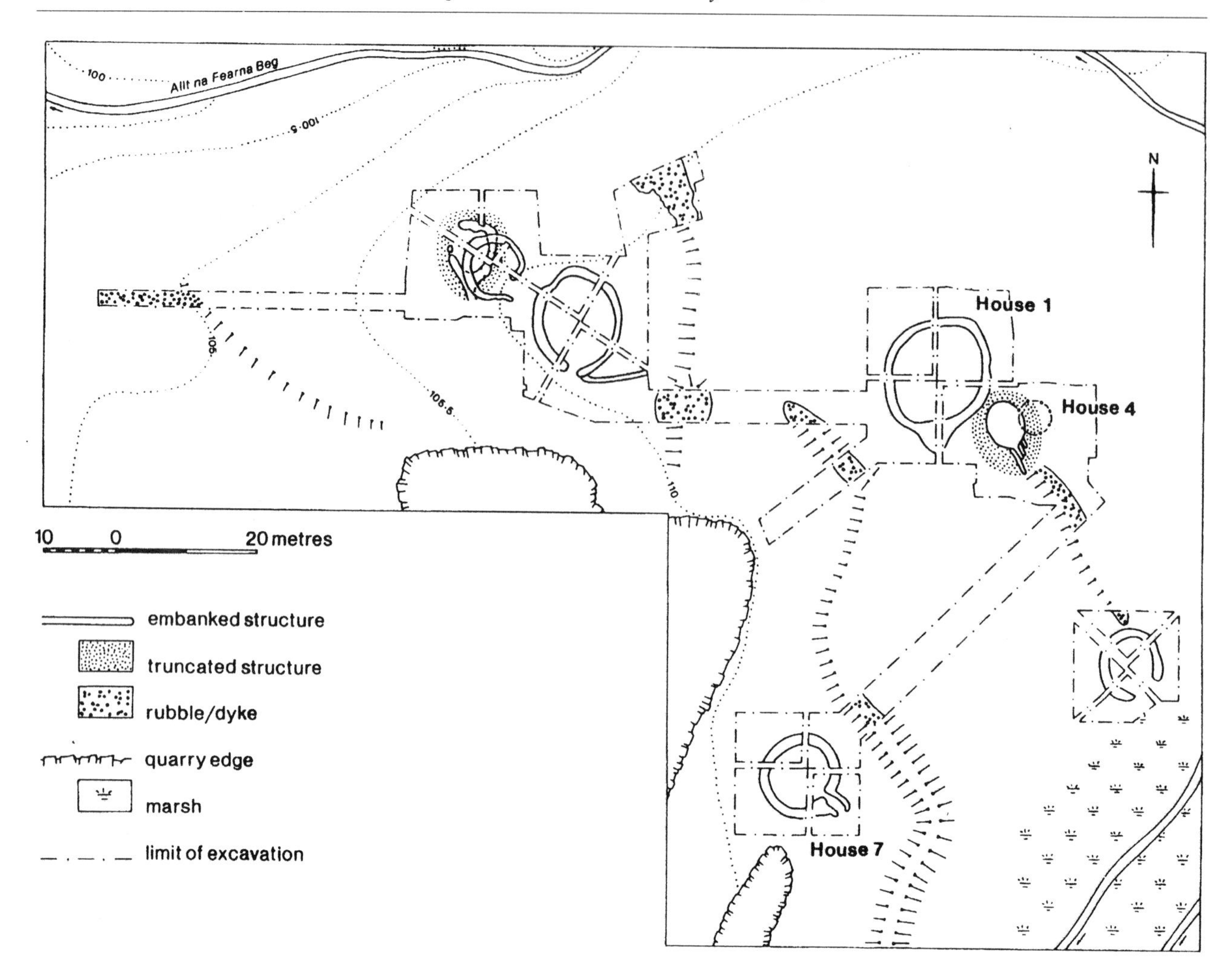

Figure 1.3: Excavated areas at the quarry indicating the round-houses referred to in the text.

Available interpretative models (Hillman 1981, 1984; G Jones 1984; M Jones 1985; and summarised by van der Veen 1992) suggest that the lack of chaff across the site is exactly what one might expect from a cereal consumer rather than a producer site. It is, however, well recognised that the very chaff components required to identify the early stages in crop processing (i.e. producer sites) in free-threshing cereals such as barley do not survive well in archaeological contexts. The chaff is both easily removed by effective winnowing and readily converted to ash if burnt (Hillman pers comm.). It is also probable that cereal threshing waste would have been of value as animal fodder or, subsequently, as a component of manure. Neither of these processes would have required the application of heat and therefore offered only slim opportunities for cereal chaff becoming charred.

Other species present such as *Chenopodium album* L.(fat hen), *Rumex acetosella* L. (sheep's sorrel) or *Spergula arvensis* L. (corn spurrey) are known from the ethnohistorical record to have been cultivated for their seed (A Steensberg pers comm, citing Hansen 1921) and potentially could have been grown at Lairg, possibly as part of a system of shifting agriculture. If this was the case, they left no appreciable archaeobotanical trace and were certainly not being stored in the two round-houses that were destroyed by fire. Likewise, it is possible, or even probable, that a percentage of the cultivated land was used for the production of root crops and greens. Members of the Compositae, Cruciferae, Polygonaceae and Chenopodiaceae all contain genera which potentially could have been utilised. Evidence to support this is again lacking but poor preservation and difficulty of identification of these plant tissues are likely to have masked their real importance.

The weed seed element (see Table 1.1) is limited, with most samples containing no more than one or two seeds. Some of the taxa have relatively large seeds (eg *Bilderdykia convolvulus* (L.) Dumort – bindweed) but the majority of the seeds are much smaller than the cereal grains which accompany them. The evidence suggests that these seeds were not just contaminants of the cleaned crop because in contexts with high cereal grain concentrations (eg, in conflagration deposits) no concomitant increase in weed seed concentrations could be detected. It therefore could

Table 1.1: Synopsis of the charred plant remains from the Bronze-Age round-house, House 4.

Latin name	Plant part	Common name	Number of items	
			Conflagration	Other
Corylus avellana L.	nut shell	hazel, cob-nut	present	present
Polygonum aviculare L.	nutlet	knotgrass	1	2
Polygonum persicaria/lapathifolium	nutlet	persicaria/pale persicaria	3	1
Polygonum sp.	nutlet	persicaria	3	1
Bilderdykia convolvulus (L.) Dumort.	nutlet	black bindweed	4	1
Rumex acetosella L.	nutlet	sheep's sorrel	1	0
Rumex cf. crispus L.	nutlet	curled dock	1	0
Chenopodium album L.	nutlet	fat hen	38	1
Atriplex hastata/patula	nutlet	orache	1	0
Chenopodium/Atriplex sp.	nutlet	goosefoots/oraches	24	0
Chenopodiaceae indet.	nutlet	goosefoot family	1	0
Stellaria media (L.) Vill.	seed	chickweed	27	2
Spergula arvensis L.	seed	corn spurrey	6	2
Ranunculus acris/repens/bulbosus	achene	buttercups	14	4
Brassica/Sinapis sp.	seed	cabbage/mustard	2	0
Calluna vulgaris (L.) Hull	flower	ling, heather	0	2
Galium aparine L.	fruit	goosegrass	1	0
Galeopsis angustifolia/tetrahit	nutlet	narrow-leaved hemp-nettle	28	0
Plantago lanceolata L.	seed	ribwort	2	0
Poa annua L.	caryopsis	annual poa	22	0
Hordeum vulgare indet.	caryopsis	barley indet.	8639	254
Hordeum vulgare (hulled)	caryopsis	hulled barley	147	14
Hordeum vulgare (naked)	caryopsis	naked barley	7177	240
Hordeum vulgare (cf. naked - 4 row lax eared)	rachis	4 row barley	0	1
Gramineae (medium seeded)	caryopsis	medium-grained grass	1	0
Gramineae (small seeded)	caryopsis	small-grained grass	8	1
Carex sp.	nutlet	sedge	5	2

be argued that the weed seed component and the cereal grains arrived on-site separately. Even so, the character of the weed assemblage is basically segetal and it is considered likely that they were originally associated with the cereal crop, possibly as straw used for kindling (but see Holden 1998 for the whole argument). In view of this, the species present can be used to reconstruct elements of the field ecology of which they were a part. Most are nitrophilous annuals, such as *Stellaria media* (L.) Vill. (chickweed) and *Polygonum* spp (knotweeds), which are rapid colonisers of disturbed ground and are therefore typical components of arable fields. A number of perennials, such as *Plantago lanceolata* L. (ribwort plantain) and *Ranunculus* spp. (buttercup – probably *R. repens* L.), were also recovered and these too show a preference for nutrient-rich soil and are quick to colonise open ground. The presence of perennial species which might be expected to be more susceptible to soil disturbance may appear anomalous but both of the above taxa readily reproduce vegetatively from cut fragments of stolon or rhizome (Sagar and Harper 1964; Harper 1957) and might therefore be expected to proliferate under an agricultural system which included a period of fallow. However, *Plantago lanceolata*, and probably also *Ranunculus* spp. can, under favourable conditions, produce seeds in the first year of growth. They could, therefore, still be expected to contribute to the seed assemblages resulting from more intensively cultivated regimes.

There are few directly comparable sites to those excavated at Lairg but it is interesting to note that the wild plant component from the Bronze Age period at Lairg is very similar to the charred assemblages recovered by van der Veen (1992) from a number of Late Bronze Age and Iron

Age sites in the north of England. Her Type A sites, in particular, yielded a similar suite of plant species to Lairg and she has interpreted these as being the product of an intensive, subsistence type of agriculture. This interpretation would be in keeping with our present thoughts regarding Lairg.

The situation can therefore be summarised thus: that barley was a staple food resource would seem evident but the charred plant remains are not able to identify whether the settlement was a cereal producer or consumer. It is possible that non-cereal crops were grown but there is no evidence to suggest that this was the case. In the absence of data to the contrary, therefore, we are left with no practical alternative than to assume that barley was the major crop. Evidence provided by the weed seeds suggests a high-nitrogen, high-disturbance field environment.

(b) Evidence from soils

Soils evidence for Bronze Age cultivation comes from fragments of land surface preserved under later structures and the sediments that accumulated in lynchets at the field boundaries. Only fragments of the Bronze Age land surface survive and therefore the extent of cultivation in the main area of excavations is best indicated by four lynchets that accumulated during this period. These lie roughly parallel, 20 to 50 m apart and up to 100 m long so the total cultivated area could therefore have been *circa* 2 ha.

In the fragments of buried land surface, the ploughsoil is a 5–15 cm deep layer of sandy loam. Analysis of thin sections reveals that this horizon has an intergrain-micro-aggregate structure with no evidence for the textural pedofeatures typically associated with cultivation (Macphail *et al* 1990). Whilst this absence of pedofeatures may be the product of the coarse texture of the soil, it is also clear that the soil structure is created from invertebrate excrement (probably *Enchytraeidae*). It could therefore have formed after the end of cultivation and destroyed any pedofeatures created by that cultivation. Indeed, the most convincing evidence that these soils were ever cultivated comes from the presence of sets of short irregular linear cuts in the subsoil surface that have probably been made by an ard.

The ard-marks are preserved in the surface of the subsoil which, in all soil profiles examined in thin section, comprises the base of a podzolic B horizon developed in the top of the indurated C horizon. The soils therefore have a shallow A–C profile and appear to be highly truncated podzols. Further evidence for soil loss is the presence of lynchets up to 0.7 m deep which occur at the downslope-side of each field. The discovery of two ploughed-out round-houses during the excavations is interpreted as evidence for intensive land-use with land occupied by derelict buildings being returned to arable cultivation. This combination of soil erosion and intensive land use is indicative of continuous arable cultivation.

Given the evidence for soil erosion (truncated profiles and lynchets) it may be assumed that maintenance of fertility, and therefore crop yield, was an issue in the Bronze Age. The chemical properties of the buried soils are strongly correlated with depth below the surface of the present day stagnopodzol soil profile. They are therefore a product of recent soil processes and are not a useful guide to past fertility. Although soils analysis has not produced any direct evidence for the past fertility of the cultivated soils, it may be assumed that they were naturally of low fertility (Futty & Towers 1982, 33). Therefore, maintenance of crop-yields must have required the use of manures or long fallow breaks. In prehistoric cultivated soils, manuring can be indicated by the presence of comminuted domestic refuse but at Lairg the evidence is equivocal. Only very low concentrations of carbonised plant remains were recovered from the sieving of bulk sediment samples and no pottery, bone, or stone artefacts were found. This situation should be interpreted as an absence of evidence because of these material types, only the stone artefacts are robust enough to survive for long in the ploughsoil; these were rare in all excavated contexts. In thin section, carbonised plant residues were only rarely encountered as large fragments but they formed a frequent component of the soil fine fraction (less than 50 μm). This could be interpreted as highly comminuted refuse but could equally derive from the burning of vegetation in other circumstances. Manuring is therefore assumed to be likely on the grounds that cultivation was maintained despite soil erosion, but cannot be demonstrated from the soils evidence.

(c) Summary synthesis of soil and plant evidence from the Bronze Age

If the soils and plant evidence is combined, this is the interpretation of Bronze Age arable agriculture:

1. The dominant crop was naked barley.
2. This crop was grown in soils with high nitrogen levels.
3. Cultivation involved the use of an ard and resulted in a high level of soil disturbance and substantial erosion.
4. The cultivated land was divided into irregular strips across the slope.

4. IRON AGE AGRICULTURE AT LAIRG

(a) Evidence from the plant remains

Deposits yielding substantial quantities of charred plant remains from later periods are much rarer than in the Bronze Age. The only structure with any appreciable carbonised plant remains from the Iron Age is a round-house (House 7 – see Table 1.2). This has produced a series of calibrated radiocarbon dates in the range 400 BC –100 AD. In contrast to those from House 4 (above) no carbonised plant remains associated with a conflagration

Table 1.2: Synopsis of the charred remains from the Iron Age round-house, House 7.

Latin name	Plant part	Common name	Number of items from all deposits
Polygonum cf. aviculare	nutlet	knotgrass	1
Chenopodium album L.	caryopsis	fat hen	2
Calluna vulgaris (L.) Hull	flower	ling, heather	8
Triticum cf. dicoccum	caryopsis	emmer wheat	1
Triticum dicoccum/spelta	caryopsis	emmer/spelt wheat	3
Hordeum vulgare indet.	caryopsis	barley indet.	100
Hordeum vulgare (hulled)	caryopsis	hulled barley	6
Hordeum vulgare (cf. hulled)	caryopsis	hulled barley	8
Hordeum vulgare L. (hulled - twisted)	caryopsis	hulled barley	1
Hordeum vulgare (cf. naked)	caryopsis	naked barley	8
Avena sp.	caryopsis	oat	2
cf. Avena sp.	caryopsis	oat	5
Gramineae (large seeded)	culm node	large-seeded grass	4
Gramineae (small seeded)	caryopsis	small-seeded grass	1

were encountered with most of the recovered material being consistent with an accumulation of grain burnt during small-scale accidents during cooking and food processing. Preservation was generally poor as evidenced by the high proportion of grain which could not be identified beyond the level of cultivated barley but hulled barley grains dominated those examples which could be identified further. Smaller quantities of wheat and oats, species which were totally absent from the earlier samples, were also recovered albeit in low numbers. It is, however, conceivable that the wheat grains represent intrusive material from earlier, truncated, Neolithic levels.

The weed seed record from this period is extremely sparse, probably as a result of poor preservation. It generally conforms to the picture presented for the earlier Bronze Age samples but the data are too poor to offer a real insight into the ecology of the fields. The presence of the flowers of *Calluna vulgaris* (L.) Hull (heather/ling) is, however, of some significance. It seems likely that heather could have been brought to the site as roofing material or fuel rather than contaminants of a cereal crop and may provide evidence for areas of heathy scrub in the vicinity of the site.

The data from this period indicate a tendency towards hulled barley in place of its naked counterpart which dominated the Bronze Age. Diversification in the use, and probably cultivation, of crop types is also indicated by the inclusion of oats and possibly wheat. Very little evidence relating to the ecology of the agricultural fields is available so more detailed reconstruction is not possible.

(b) Evidence from soils

Evidence from soils for Iron Age cultivation comes from two sources: a single example of a soil buried under an Iron Age round-house (House 7) and large areas of ridged soils preserved directly under shallow blanket peat. House 7 is dated by radiocarbon to the period 400 BC to AD 100 and the blanket peat has basal dates in the later first millennium AD.

The soil buried by House 7 is similar in many respects to the Bronze Age soils. The ploughsoil is somewhat deeper (20 cm) but it has the same sandy loam texture and intergrain-microaggregate structure and lacks pedofeatures attributable to cultivation. Differences are apparent in the abundance of two components: fine charcoal and fragments of indurated till. Fine charcoal is rare compared with Bronze Age ploughsoils; if the charcoal is derived from manuring then it may indicate less, or possibly a different type of manure being applied. Only single fragments of indurated till were noted in the Bronze Age soils but in this soil frequent examples are present towards the base of the ploughsoil. This suggests that the cultivation implement was actually penetrating the indurated layer rather than running over the top of it. Infrequent but parallel ard-marks up to 5 m long were recorded at the base of the ploughsoil under House 7. Given that the dating of the overlying house allows an Iron Age date for this cultivation, it is suggested that the change in the nature of the ard cultivation reflects the introduction of iron ard shares (replacing the earlier wooden shares), capable of much greater ground penetration.

Table 1.3: Summary of the evidence realting to cultivation techniques used at Lairg.

Bronze Age	Iron Age	Inferences
No evidence for narrow ridges	Narrow ridges	Change in agricultural techniques
Ard marks with single preferred orientation	Ard marks indicate cross-ploughing	Possibly linked to the use of an ard to bring fields out of fallow in the Iron Age
Short, irregular ard marks	Long, narrow ard marks	Use of lighter ards in Bronze Age and then iron plough share and rip ards in the Iron Age possibly to break fallow
Rare fragments of indurated subsoil in agricultural soil	Common fragments of indurated subsoil in base of agricultural soil	Ploughing in the Iron Age being used to break up the indurated subsoil
Fields *circa* 2 ha	Ridged plots up to 0.25 ha	Changes in land division
Higher soil charcoal concentrations	Lower soil charcoal concentrations	The use of domestic hearth debris as a manure was more prevalent in the Bronze Age
Soil movement and the formation of lynchets	No evidence for large scale soil movement	Erosion in the Bronze Age is a significant factor whereas there is little evidence for it in the Iron Age
Ploughing up of earlier roundhouses	Ploughing round earlier roundhouses	Good land was at a premium in the Bronze Age but a less intensive form of agriculture in the Iron Age was used
Shallow soil *circa* 5-15 cm	Deeper soil *circa* 20 cm	Greater quantities of the subsoil were being worked into the agricultural soils in the Iron Age
Naked barley is dominant crop	Hulled barley is dominant crop	The major cereal type changes between the Bronze Age and the Iron Age
No evidence for wheat or oat	Oat and possibly also wheat present	New crop species are introduced in the Iron Age
Weed seeds indicative of high-nitrate, high-disturbance environment	Very few weeds seeds present because of poor preservation	Manuring and intensive tillage were practised in the Bronze Age

Ridged soils of late Iron Age date (first millennium AD) were found surrounding the earlier, Bronze Age buildings and a series of test pits revealed them extending over a wide area. Ridges are set 1.0 to 1.5 m apart and groups with the same orientation cover areas up to 70 x 40 m in extent. They do not cross existing boundaries, respecting the Bronze Age fields and round-houses, and more than one group of ridges occurs within each of the older fields. At the base of the Iron Age ploughsoil there are well defined ard-marks forming two parallel sets at right angles. One set is parallel to the overlying ridge and furrow. The ridges are now only 10 cm high with a total topsoil depth of 20–25 cm.

Cross-ploughing has been frequently recorded in the past, including other narrow ridge sites (for example at Rudchester; Gillam *et al* 1973). It has been interpreted as the product of the so called 'rip ard' (Reynolds 1981, 102 –104), designed to break-in fallow land prior to cultivation. The common orientation of ridges and ard-marks at Lairg is consistent with the initial cutting of turf for the creation of turf ridges. There is no evidence for soil erosion during ridge cultivation or the widespread levelling of existing buildings. This may be interpreted as the result of turf ridging which reduces the bare soil surface available for erosion or perhaps indicates that the ridge cultivation was of limited duration. The survival of derelict buildings suggests a low intensity agricultural regime, in contrast with the Bronze Age evidence for the demolition of buildings and reinstatement of agricultural land.

(c) Summary synthesis of soil and plant evidence from the Iron Age

If the soils and plant evidence is combined, this is the interpretation of Iron Age arable agriculture:

1. The dominant crop is hulled barley but naked barley is also present together with lesser quantities of oats and possibly also wheat.
2. Cultivation involved the breaking of fallow with an ard and then the creating of narrow turf ridges. This did not lead to significant erosion.
3. The ard was capable of consistently penetrating the subsoil.

Figure 1.4: Ard-marks resulting from cross-ploughing in the area immediately to the east of House 1. These marks were overlain by a ridged cultivation soil. The ridges were aligned parallel to the SW-NE orientated ards marks.

4. Evidence for the use of manure is insubstantial but the data seem to suggest lower applications of domestic refuse on the fields.

5. DISCUSSION

The joint use of data on soils and plant remains has allowed the resolution of some taphonomic problems that affect one or other of the source materials. Several strands of evidence presented here suggest that substantial changes did occur in arable agriculture between the second millennium BC and the first millennium AD at Lairg. These are apparent in crops, tools and cultivation practice as summarised in Table 1.3. Central to these changes appears to be the shift from flat to ridged-field cultivation. However, looking at the wider perspective we find that the evidence need not be explained in terms of changes in cultivation techniques at all and some alternative explanations are suggested. In spite of this, agricultural change remains a strong possibility and the feasibility of several possible explanatory hypotheses are also discussed.

(a) Is the observed change in cultivation practice "real" or "apparent"?

Differential survival

The appearance of narrow ridging associated with Iron Age settlement at Lairg appears to suggest a substantial change in arable agricultural techniques. However, present evidence from other excavated sites in Britain demonstrates that narrow ridged fields were not an Iron Age innovation. At North Mains, in central Scotland, narrow ridges were found beneath a large Late Neolithic mound (Barclay 1989). They were also recorded from Glenree, County Mayo, Ireland, beneath peat and date to the second millennium BC (Herity 1981). The technique therefore appears to have been used in the north and west of the British Isles from at least as early as the beginning of the second millennium BC.

Agricultural features from earlier periods are susceptible to destruction by subsequent agricultural activity. Most of the surviving examples of prehistoric ridged fields on the Anglo-Scottish Border, for example, are found above 300 m OD (Topping 1989) and are therefore above the limit where subsequent Medieval cultivation would have destroyed them. However, examples such as those from Wallsend at only 15 m OD (Topping 1989) demonstrate that they were not just a highland phenomenon and tend to support the assertion that most lowland examples have been destroyed. At Lairg it is evident that much of the area has been intensively cultivated since the Bronze Age and so it is difficult to demonstrate that the apparent abundance of ridged fields in the Iron Age is a genuine reflection of changing agricultural practices. However, the Iron Age ridges at Lairg are also associated with regular cross-ploughing as demonstrated by ard-marks in the subsoil. These should be less susceptible to destruction than the ridges themselves so it is possible that the apparent absence of cross-ploughing in the Bronze Age is a more reliable indicator of a change in cultivation techniques. The surviving areas of Bronze Age ard-marks contain only one preferred orientation at most.

The identification and excavation of fields at Lairg, last cultivated in the Bronze Age, would provide a more complete record of the cultivation regime and therefore confirm or refute the apparent absence of ridging.

Spatial bias

It is possible that the two methods of cultivation, flat fields and ridged fields, are contemporary, complementary techniques. The first type, as observed in the fields associated with the Bronze Age settlements at Lairg, representing an intensive "infield" technique and the other, represented by ridged fields an extensive "outfield" technique. If this is the case then the sequence observed at Lairg may be an artefact of the spatial distribution of fields around the different settlements. The fields close to the Bronze Age round-houses could represent their infields whereas the later ridged fields could represent the outfields of an as yet unexcavated Iron Age settlement located some distance away. It may be noted that a number of round-houses were excavated that were broadly contemporary and adjacent to the Bronze Age fields whereas only one Iron Age house was found adjacent to the Iron Age ridged fields.

Given the long recorded chronology for ridged fields, it seems possible that both ridged and flat field cultivation were used side by side throughout prehistory, performing different functions. Further data with which to test this proposition would come from the excavation of an Iron Age settlement and its immediate surroundings. A few Iron Age settlements have been identified at Lairg and they could be investigated.

(b) If the changes in cultivation practice are real then how could they be explained?

Climatic change

It has already been proposed by other writers that ridge cultivation is designed to alter soil microclimate by improvement of drainage and soil depth. It is therefore possible that the adoption of ridging in the Iron Age could have been a response to the deteriorating climatic conditions. Certainly, in areas which were probably marginal with respect to the cultivation of cereals (of which Lairg is one) even a minor shift in weather patterns could have necessitated the adoption of different agricultural practices. Although much of the evidence for increased rainfall at this time in northern Scotland is, at present, tentative the data from Lairg would be compatible with such an explanation.

Change in crop type

At Lairg, and elsewhere in upland Scotland, there is a shift from naked barley in the Bronze Age to hulled barley and other cereals in the Iron Age. Whilst, in principle, a new crop might require different cultivation methods, the recorded shift from naked to hulled barley can hardly have merited such a change. It is possible that both changes in crop and cultivation technique reflect climatic change but this would be autocorellation rather than a causal link.

Demographic change

The link between narrow ridge cultivation and cross-ploughing has been interpreted as evidence that the technique involved the cutting of a sod to create turf ridges. It can be argued that an increase in population would require the taking in of more marginal land for agriculture. If ridging was the most effective way of making this land suitable for cultivation by increasing soil depth and improving drainage, possibly as part of a fallow regime, then an increase in area of ridged fields could be predicted. The more fertile areas would, however, probably not be cultivated by the use of ridges so giving rise to an infield-outfield system of agriculture. Alternatively if ridging represents the extensive use of land with long periods of fallow then an increase in ridged fields could reflect reduced population pressure and a switch from intensive to extensive farming methods. It is not possible with the available evidence to determine which of these interpretations is correct at Lairg (if either of them are relevant at all).

6. CONCLUSION: TAPHONOMY AND INTERPRETATION

This paper has taken two lines of evidence, soil and charred plant remains, and brought them to bear on the nature of the prehistoric agriculture at Lairg. Each source of evidence has its own suite of taphonomic problems yet in a number of important areas they have proved complementary. At a very basic level, for example, the soil evidence and field archaeology very clearly indicate that agriculture was being practised in the vicinity of the excavated settlement sites but they offer no insight into the crops grown. The plant remains on the other hand, have identified the crops that were being stored and presumably consumed but these need not have been grown close to the settlement. There were no indicators for the early part of the crop processing sequence and importation of the cleaned cereal grain from elsewhere is a distinct possibility. By taking the two forms of evidence together we have a much stronger indication that it was in fact barley that was being grown in fields around the settlement. At a more detailed level, aspects of soil fertility and manuring became important. The soil evidence shows that the local soils were inherently infertile and although there is some tentative evidence for manuring in the form of comminuted charcoal fragments other events could equally well explain our observations. The plant remains, however, show quite clearly that nitrophilous species were present indicating that they had come from a high-disturbance, high-nitrogen environment. Since these are thought to have arrived on site along with the crops they strongly suggest that manuring of the fields must have been practised, at least during the Bronze Age.

Three specific research questions were asked of the data (above). We feel that we have answered the first of these and have demonstrated that, from the samples available to us, there is an apparent difference between the cultivation techniques used in the Bronze Age and the Iron Age (summarised in Table 1.3). The second question, however, asks whether these differences are real or not. This is more difficult for us to answer at present. We find that the data are wanting in several important areas, some of which are clearly the result of taphonomic processes which have acted on our source material. Of these, later agricultural disturbance, weathering of the deposits and continuing pedogenesis are particularly important. More area excavations and subsequent soil and archaeobotanical analyses of specific features would be especially useful. These would include area excavations of fields as well as their associated settlement sites since it is the spatial relationship between these that gives rise to some of the most serious interpretative problems. Fields last cultivated in the Bronze Age together with Iron Age settlements and their infields would yield the appropriate data. The use of other lines of investigation such as bone, insect and local soil pollen analyses have been hampered by poor preservation and others, such as regional pollen analysis, have yet to be completed. These may also eventually add to the overall picture.

With respect to the third of our research questions, why and how did the apparent changes in cultivation occur, much of this remains speculation although we have offered some potential explanations of the data. This is largely because we are still unsure whether we are dealing with a real change but present interpretative frameworks are barely adequate. The inclusion of other sources of new data directly focusing on the nature of prehistoric cultivation should, however, help to produce more robust hypotheses in the long term.

ACKNOWLEDGEMENTS

The authors would like to thank Rod McCullagh, Dorothy Rankin, Coralie Mills and Ann MacSween for their useful discussions and help in the production of this paper. Funding for both the excavation and post–excavation analysis was provided by Historic Scotland. Much of the soil analysis reported here was undertaken by Tim Acott as part of a SERC/Historic Scotland funded CASE studentship at the University of Stirling. The figures were prepared by Christina Unwin and Tanya O'Sullivan.

BIBLIOGRAPHY

Barclay, G.J. 1989 The cultivation remains beneath the North Mains, Strathallan, barrow, *Proc Soc Antiq Scot*, 119, 59–62.

Futty, D.W. & Towers, W. 1982 *Soil and land capability for agriculture. Northern Scotland.* Aberdeen, The Macaulay Institute for Soil Research.

Gillam, J.P., Harrison, R.M. & Newman, T.G. 1973 Interim report on the excavations at Rudchester 1972, *Archaeologia Aeliana, 5th series*, 1, 81–85.

Hansen, H.P. 1921 *Fra gamle dage 1.* Herning: Forfatterens Forlag.

Herity, M. 1981 A Bronze Age farmstead at Glenree, County Mayo, *Popular Archaeol*, 2, 36–7.

Hillman, G. 1981 Reconstructing crop husbandry practices from charred remains of crops. In: Mercer, R. (ed) *Farming Practice in British Prehistory.* Edinburgh: Edinburgh University Press, 123–162.

Hillman, G. 1984 The interpretation of archaeological plant remains: the application of ethnographic models from Turkey, in van Zeist, W. & Casparie, W.A. (eds) *Plants and ancient man*, 1–41. Rotterdam: Balkema.

Holden, T.G. 1998 Charred plant remains . In: McCullagh, R.P.J. and Tipping, R. (eds) *The Lairg Project 1988–1996: The Evolution of an archaeological landscape in Northern Scotland* 165–172. Edinburgh: Scottish Trust for Archaeological Research.

Jones, G.E.M. 1984 Interpretation of archaeological plant remains: Ethnographic models from Greece. In: van Zeist, W. & Casparie, W.A. (eds) *Plants and ancient man*, 43–61. Rotterdam: Balkema.

Jones, M. 1985 Archaeobotany beyond subsistence reconstruction in Barker, G.W. & Gamble, C. (eds), *Beyond domestication in prehistoric Europe*, 107–128. London: Academic Press.

Macphail, R.I., Courty, M.A. & Gebhardt, A. 1990 Soil micromorphological evidence of agriculture in north-west Europe, *World Archaeology*, 22, 53–69.

McCullagh, R.P.J. 1992 The Excavations at Lairg, *Current Archaeology*, 133, 455–459.

McCullagh, R.P.J. 1993 An interim report on the results of the Lairg Project 1988–1992, *Northern Studies*, 30, 34–52.

McCullagh, R.P.J. & Tipping, R. 1998 *The Lairg Project 1988–1996: The Evolution of and archaeological landscape in Northern Scotland* 165–172. Edinburgh: Scottish Trust for Archaeological Research.

Reynolds, P. 1981 Deadstock and Livestock. In: Mercer, R. (ed) *Farming Practice in British Prehistory*, 97–122. Edinburgh: Edinburgh University Press.

Topping, P. 1989 Early Cultivation in Northumberland and The Borders, *Proc Prehist Soc*, 55 (1989), 161–179.

van der Veen, M. 1992 *Crop husbandry regimes: An archaeobotanical study of farming in northern England 1000 BC – AD 500.* Sheffield Archaeological Monographs 3, Department of Archaeology and Prehistory, University of Sheffield.

2. Palynological taphonomy in understanding vegetation history and human impact in the Lairg area, Sutherland.

Melanie Smith

SUMMARY

To interpret any fossil assemblage an understanding of the taphonomy of the material is critical. In palynological investigations associated with archaeological research, it is important to appreciate the taphonomic processes involved. It is important to design a sampling strategy encompassing these processes to answer particular questions posed by the research project.

The palynological investigations for the Lairg project offered an opportunity to construct a strategy for site selection and sampling which would meet the criteria of the archaeological and ecological aims of the project.

The research area of Lairg, Sutherland, is predominantly an open moorland landscape. The region has a long record of human occupation, with recently excavated archaeological monuments from the Neolithic, Bronze Age, Iron Age, Norse and Medieval periods. Palaeoecological investigations have been undertaken to establish the vegetation history and extent of human impact on the landscape.

The sampling strategy, following the model of Jacobsen and Bradshaw (1981), was designed to obtain evidence representing local to regional vegetation cover. The relative success of using the model in selection of sites is assessed in discussion of the sites found and sampled for the palynological research.

1. INTRODUCTION

Answering palynological questions generated by archaeological investigations requires the selection of sites suited to those questions. In recent years, much research has been carried out into the factors influencing recruitment of pollen to a site, such as basin size, sediment type and pollen type (Oldfield 1970, Bradshaw 1981, Jacobsen and Bradshaw 1981). As a result of this work, theoretical models have been produced which can be used for a more critical testing of hypotheses (Webb *et al.* 1978).

These models have rarely been applied to site selection, with objectives other than palaeoecological or palaeoclimatic applications (Jacobsen and Bradshaw 1981). This paper aims to show how such a research strategy can be constructed to address some of the issues and questions being raised by archaeological investigations. The excavations at Lairg, Sutherland form the focus of the palynological site selection strategy.

2. BACKGROUND TO THE LAIRG PROJECT

The research area of Lairg, Sutherland (Figure 2.1) is predominantly an open moorland landscape, with valleys containing *Betula/Alnus* and *Quercus* woodland. The area has a long record of human occupation, from the Neolithic to the present (McCullagh 1992). Archaeological excavations from 1988–1991, combined with an intensive radiocarbon dating programme, have confirmed that phases of intensive settlement activity, for example through the mid-late Bronze Age and again in the post-Medieval period, seem to have been separated by periods with little or no such occupation (McCullagh 1992).

The significance of these periods of little of no occupation, in terms of regional landscape change, cannot be ascertained from the archaeological record alone. There were considerable constraints on the spatial scale of the archaeological work which are chiefly related to cost and practicality. The archaeological excavation work was carried out in response to the proposed widening of the A836 road south of Lairg (Figure 2.2) which obviously constrained the survey to the corridor of road widening. Cost allowed only two seasons of excavation work. Therefore other, complementary and research, methods were applied.

Excavations concentrated on the area around a small

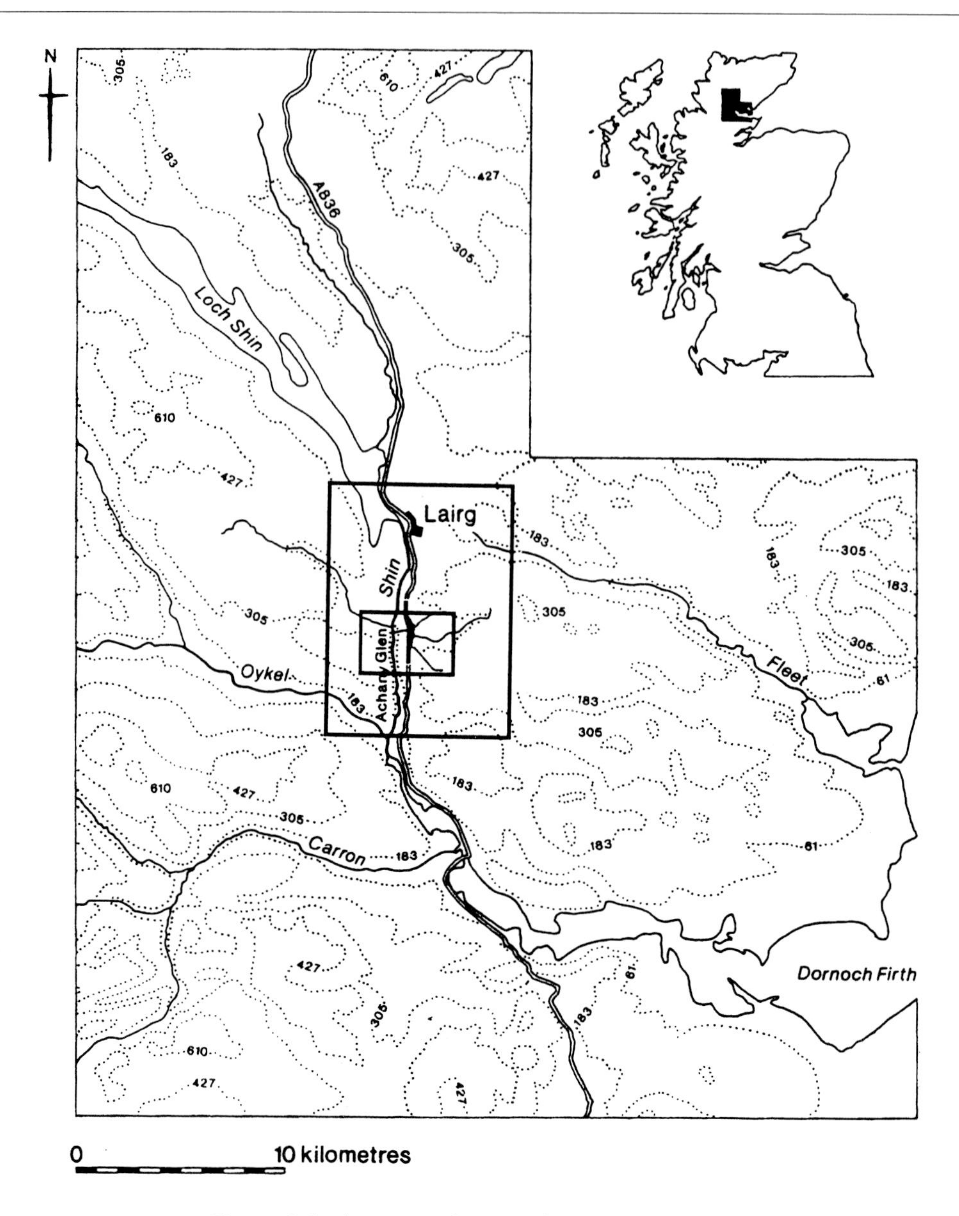

Figure 2.1: the research area of Lairg, Sutherland.

quarry adjacent to the road (Figure 2.2). Although much detailed information has been derived from these investigations it is, nonetheless, limited to a very local area around the site. For example, only four hut circles have been excavated. This means that the wider settlement history of the area can only be generally inferred from the archaeological research. The main conclusions from the archaeological work are that there is no direct structural evidence relating to the Mesolithic period but there is some evidence of Neolithic people in the survey area and more Neolithic activity in the area of The Ord (see Figure 2.2). Structures and agricultural activity associated with the Bronze Age have been found. This evidence, plus the radiocarbon results, suggests continuity of settlement from ca. 4300 cal. BP through to ca. 3000 cal. BP, after which there is a break in the evidence of settlement until ca. 2450 cal. BP. After this phase there is a return to relatively intensive human activity around the site from the Iron Age to the present (McCullagh pers.comm.).

The marked break in settlement at ca. 3000 cal. BP has been the centre of much discussion. Burgess (1990) and Barber (pers. comm.) have put forward the hypothesis that the apparent abandonment of land can be linked to the eruption of the volcano Hekla at ca. 3000 cal. BP, the so called 'Hekla 3' eruption. This hypothesis has been partly supported by the work of Baillie (1991) on tree rings in Irish bog oaks which suggests a significant climatic deterioration at around this time. In an already marginal environment, a marked climatic shift could be sufficient to force people to move from land that could no longer be productive. Burgess (1990) has linked the effect of volcanic eruptions to fluctuations in human populations

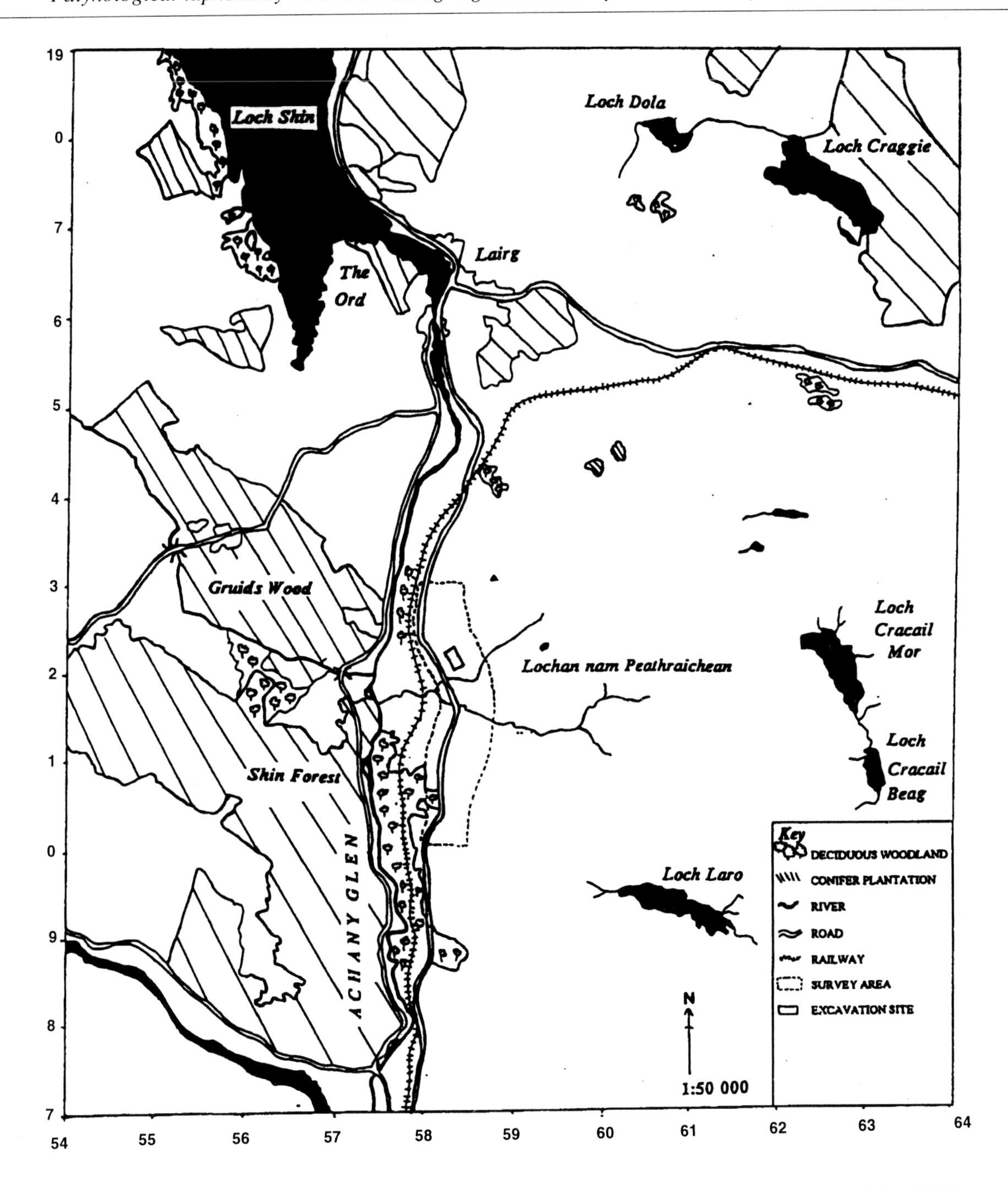

Figure 2.2: the area under archaeological investigation due to the proposed widening of the A836.

throughout the last 7000 – 8000 years. Evidence that the effects of volcanic eruptions in Iceland did reach northern Scotland can be demonstrated by the presence of thin layers of tephra in sediments in Scotland (Dugmore 1989).

The archaeological investigations have limitations and cannot be used on their own to investigate the hypothesis that fluctuations in climate, linked to volcanic eruptions, are associated with fluctuations in land use and settlement density in the Lairg area. however, research into the environmental changes which have taken place here has much to contribute to a wider interpretation of the landscape and human impact history.

The environmental investigations have concentrated upon palynological techniques to reconstruct the vegetation history. Pollen analysis can be used to address a number of key questions which cannot be answered by the archaeological research:

1 When and why was there a sudden abandonment of land c.3000 cal.BP?

The palynological evidence can indicate much more clearly the temporal and spatial scale of any such abandonment as well as the nature of landuse at the time. Pollen analysis may also identify whether there was a widespread abandonment of the land which reached beyond the immediate area around Lairg. The archaeological investigations are based on a small sample, with difficulties in dating due to truncated stratigraphy.

2 Were there phases of less intensive occupation? Do these relate to only a very local area or were people living at the same population density but a few hundred metres away?

As in question 1, palynology can provide greater insight into the spatial and temporal scales involved, providing that a well devised sampling strategy has been adopted, taking into account the factors influencing pollen taphonomy.

3 Were people able to live at higher altitude than subsequently because of a more equable climate?

An understanding of the vegetation history should be able to identify the impact of people on the landscape at different altitudes both temporally and spatially providing that the pollen recruited to the sites selected is representative of the temporal and spatial scales being investigated.

These three questions cover the main issues which can be addressed by the palynological investigations. More specific and detailed questions are beyond the scope of this paper and are generally covered by the issues raised in the questions listed above.

3. THE THEORETICAL BASIS OF POLLEN RECRUITMENT

The palynological investigations at Lairg (Figure 2.3) have utilised a sampling strategy derived from theoretical and empirical work on pollen taphonomy. Taphonomy can be defined as "investigating the way in which pollen arrives at and is preserved in various sites" (Birks and Birks, 1980).

Several aspects are important in the recruitment of pollen to a site. These factors are complex but can be broadly categorised as:

1. Those factors related to properties of the parent plant, such as whether a plant species is insect- or wind-pollinated, or those relating to factors influencing the settling velocity of pollen grains (Prentice, 1988).
2. Those properties relevant to the size of the site, and therefore the influence this will have on recruitment of pollen (Bonny, 1978).
3. The nature of the site in which the pollen is preserved, whether a lake, peat or soil. This has a direct influence on the stratigraphic integrity of the pollen assemblage, e.g. through reworking subsequent to deposition or through soil erosion (Birks and Birks, 1982). In lakes, reworking can be due to annual turnover or the effect of wind fetch across the water surface (Bonny, 1978).
4. The state of preservation of the pollen will introduce biases in pollen representation. The fossil assemblages, therefore, may differ in composition from the assemblages as originally deposited (Konigsson, 1969).

Given the requirements of the hypotheses needing to be tested at Lairg, emphasis in this paper is on those factors important in assessing the spatial resolution of pollen events, namely points two and three. However, all the taphonomic factors interact, and need to be considered in an integrated model of pollen recruitment.

4. DEFINITION OF JACOBSEN AND BRADSHAW'S MODEL (1981)

Figure 2.4 is adapted from Jacobsen and Bradshaw's model to show the sites selected for research in relation to the original parameters of the model. AG1, 2 and 3 refer to the sites in Achany Glen that were used in the Lairg study. From this diagram the main principles of the model can be ascertained. Janssen (1966) defines the scale of pollen recruitment to a site as local, extra-local and regional, and relates this to the size of the site itself. Tauber (1965, 1977), presents a model defining the various components contributing to pollen transport to a site (Table 2.1).

Table 2.1: Pollen components (after Tauber 1965, 1977)

Code	Mode of Transport
Cw	Streams and Surface runoff
Ct	Trunk space
Cc	Above the canopy
Cr	Rainfall

Jacobsen and Bradshaw (1981), took elements of both Janssen's and Tauber's models and, in addition, included a further component, gravity (Cg), which describes the input of pollen from plants on the edge and hanging over the basin (Jacobsen and Bradshaw, 1981). Figure 2.2 shows how basin size is related to both area of origin of the pollen and mode of transport of the pollen to the site.

"This model will serve as a guide for estimating the form of the pollen source area for basins that have no inflowing streams...We define local pollen as origin-

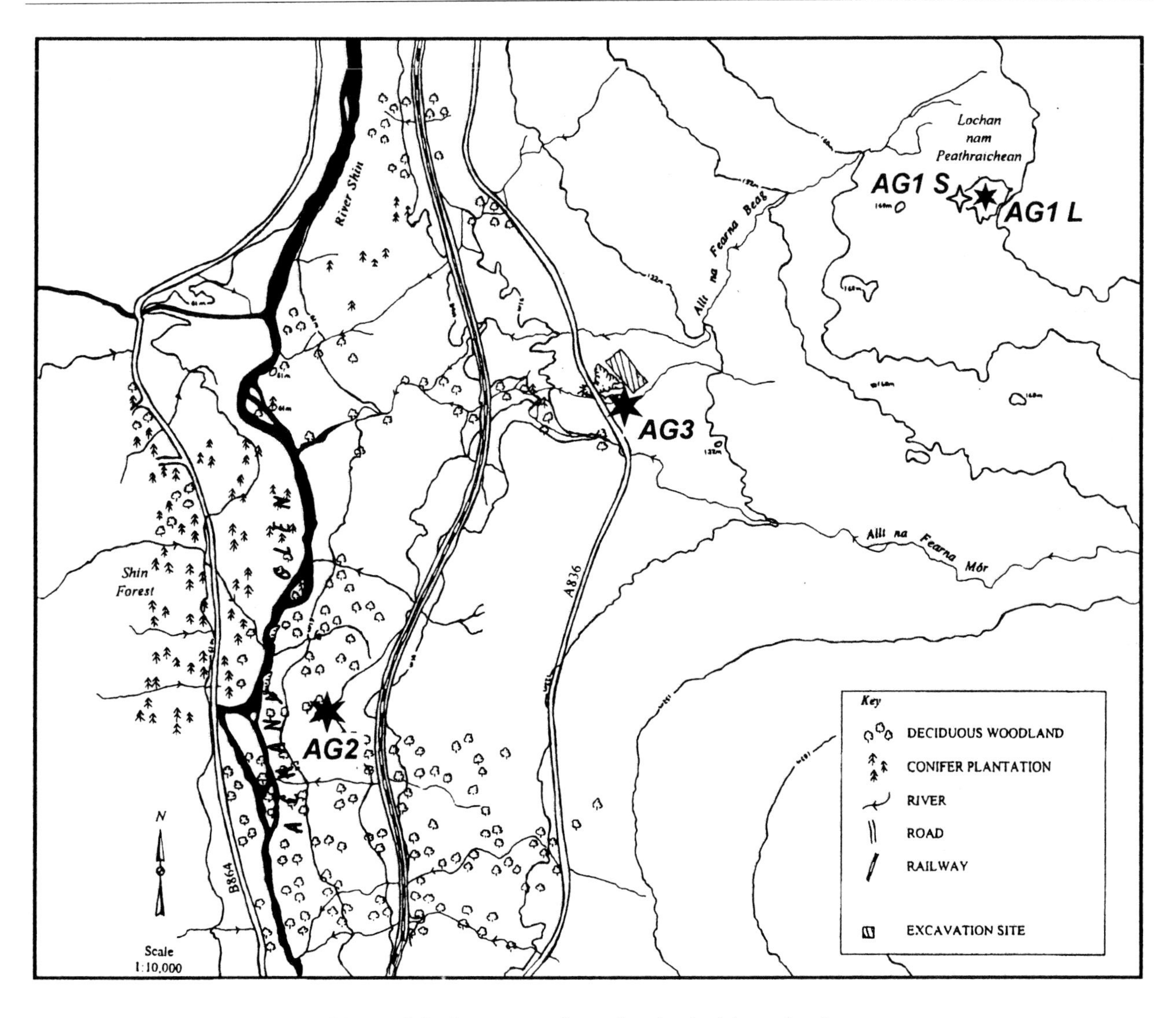

Figure 2.3: the area under palynological investigation.

ating from plants growing within 20m of the edge of the sampling basin, extralocal pollen as from plants growing between 20 and several hundred meters of the basin, and regional pollen as derived from plants at greater distances.....This model is only a general guide, and local anomalies of hydrology or topography will influence the outlined predictions." (Jacobsen and Bradshaw, 1981).

There still remains the very relevant problem, identified by Oldfield (1970), of distinguishing between many individuals at some distance from the site as opposed to a few individuals close to the site. To obviate this problem in the current research a range of sites was selected which fulfilled the criteria for each type of basin/pollen recruitment area defined by the model. By doing this within a spatially constrained field area, the interplay of different recruitment areas would be enhanced. The Lairg area offered a good opportunity to find such a range of sites, given the considerable choice of both peat bogs and lake basins. By applying the model in this way, the main excavation site of the archaeological investigations, representing only a very localised history of human settlement, was used as a focus for the site selection strategy.

5. THE METHODOLOGY BEHIND THE SELECTION OF SITES

The model of Jacobsen and Bradshaw (1981) (see Figure 2.4) was used as a basis for selection of sites to fulfil the requirements of the project in addressing the issues and questions as previously discussed. A detailed proforma was drawn up defining a set of criteria used to eliminate and ultimately select sites. Details such as location, aspect, altitude, the type and size of basin, presence of inflowing and outflowing streams, distance from the excavation site and the nature of the surrounding vegetation were recorded (Appendix 2.1). The size of the basin, sediment type and distance from the known archaeology are the key criteria.

The size of the site can be related directly to the model of pollen recruitment (Jacobsen and Bradshaw, 1981). The size ranges for each category of site are given in Table 2.2.

Table 2.2: Size ranges for each site category.

Site Category	Size Range
Local	< 30 m diameter
Extra – local	30 – 150 m diameter
Regional	> 150 m diameter

It was important to select a site from each size category. Of almost equal significance was the type of sediment. A lake site was the preferred regional-scale site as lake muds have a relatively undisturbed, continuous record (Pennington, 1979; Bell and Walker, 1992), although a contrast between the palynological records from lake muds and peats would also add further value to the investigation.

The distance from the archaeological excavation site was important in terms of the model as it provided the focus from which to assess local to regional pollen recruitment to a site. The local scale site would need to be very close (preferably within 40–50 metres) to the excavations if it was to show pollen representing human impact that could be related to the excavated settlement history. All of the sites had to be representative of the pollen recruitment of the Lairg area at the relevant period in prehistory and not be disjointed in either time or space.

The presence of inflowing and outflowing streams is important in terms of the Cw component of the model. Ideally a site should have no inflowing or outflowing streams as this introduces a bias to pollen recruitment to the site. However, the presence of an inflowing stream has greater significance, as this can bring in pollen from a very wide catchment area. For example, Loch Shin has inflowing streams which ultimately drain a catchment area extending over much of northern Scotland. The pollen signal from Loch Shin would be hard to interpret in relation to any particular archaeological site because of its large pollen recruitment area.

The altitude and exposure of the site were recorded to gain further information on the environment around the site. Both are particularly relevant to the vegetation types growing around the site. For example, the upper limits of natural Scots pine growth is at around 300m (Bridge *et al* 1990). Jacobsen and Bradshaw's model was derived from data collected at sites in southern Britain and they did not consider the exposure of those sites. However, exposure and consequent effects of wind across a site could well be an important factor when considering pollen dispersal in northern Scotland.

The area considered in detail was 6 x 8 km around Lairg (Figure 2.3). Possible sites in the wider area were also considered, particularly those which may have fulfilled the criteria for a regional scale site, for example Loch Shin. The area is characterised by extensive blanket peat, valley bogs and numerous lochs. This widened the range of potential sites, making the criteria particularly useful in defining the most appropriate type of site. By using a well defined set of criteria, before beginning to sample any one site, it was possible to reject large numbers of sites which did not satisfy the full criteria of the model and requirements of the project. This has ensured the research to be scientifically objective as well as efficient in use of time and resources.

Those sites fitting the primary criteria were cored using a gouge auger. This was primarily to establish the depth of the site and the nature of the sediment. The depth of any site needed to be great enough to obtain a reasonable temporal resolution, so any site with less than two metres of sediment was generally discarded. The nature of the sediment at the base of the core was of value as it gave an insight into the origins of the site, for example, a tripartite Late Glacial sequence gives a very useful indication of the age of the site (Lowe and Walker, 1984).

6. DISCUSSION OF SITES SELECTED

Sites were prospected during two seasons of fieldwork in 1992/93. Despite the numbers of potential sites, it still proved difficult to identify any exactly meeting the criteria. The following three sites in Achany Glen were selected and sampled (Figure 2.3).

(a). A local site – Achany Glen Site 3 (AG3) National grid reference: NC 583 018

The site is a small valley peat bog, ca. 40m in diameter at the widest point. It lies immediately adjacent to the archaeological excavations. It therefore fulfills, as closely as possible, the criterion for a local scale site: pollen recruitment to the sediments should reflect nearby vegetation changes which can be related to the known human settlement sequence of the archaeological investigations.

There are no inflowing or outflowing streams which influence the basin directly, although there is a small stream on the northern edge of the basin. The sediments are entirely peat which is an ideal medium for preservation of fossil evidence (Bell and Walker, 1992), therefore a detailed and well preserved pollen record would be expected from the sediments sampled at AG3 unless they had dried out at some time. Peats are also known to preserve evidence of human activity in the form of inwashed layers of minerogenic material (Dorfler, 1992). This feature is of particular relevance to AG3 given the

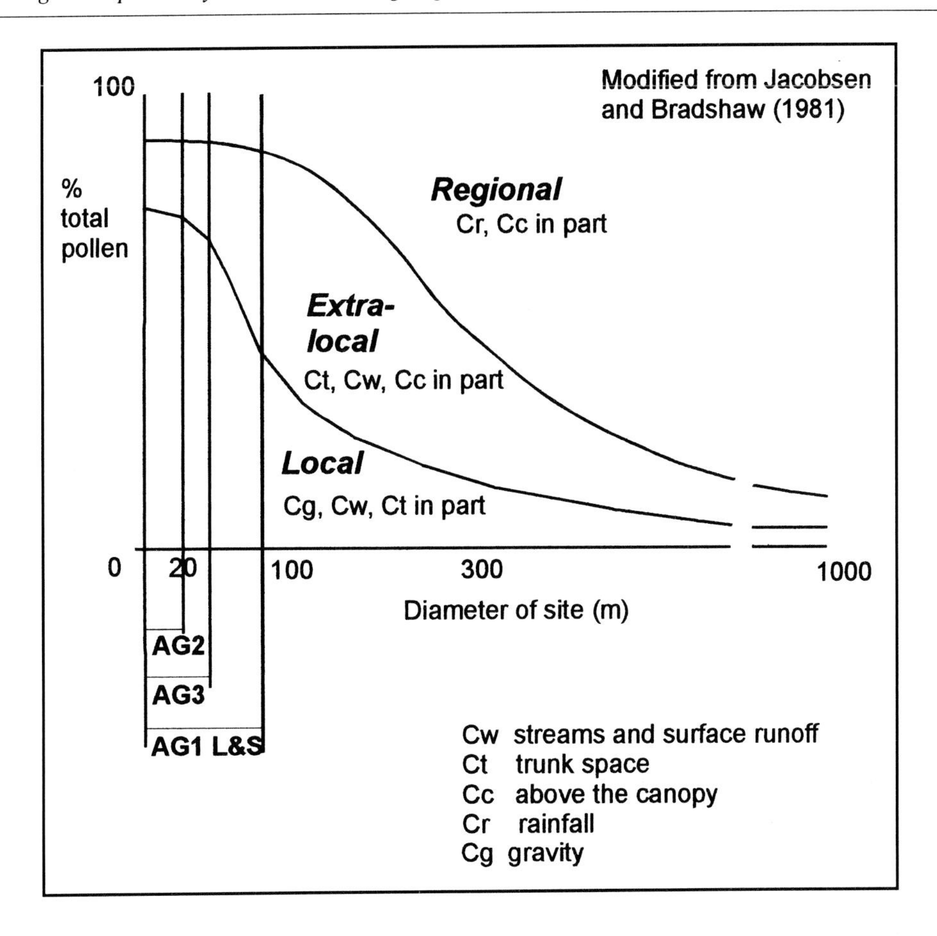

Figure 2.4: Pollen taphonomic pathways (after Jacobsen and Bradshaw, 1981).

proximity to a known area of human activity. Peats have also been the medium in which tephra has been found in Scotland (Dugmore, 1989), material providing independent dating of sediments. The sediments have a total depth of 276cm, sufficient for reasonable temporal resolution.

(b). A regional site – Achany Glen Site 1 Shore (AG1S) and Achany Glen Site 1 Lochan (AG1L). National grid reference: NC 592 023

The site, Lochan nam Peatriarchean, is located on a plateau to the east of the excavation area. It is a small lochan, 100 metres across at the widest point, and therefore fulfills the size criterion for an extra local site. However, the effects of exposure on the plateau are likely to be especially important at this site. This factor is considered critical in deciding whether the lochan basin will be representative of extra-local or regional pollen. Given the considerable exposure and after considering other possible regional sites, including Loch Dola to the north and Loch Craccail Mor to the East, it was decided to treat AG1 as a regional site. It is important both in terms of the model and the objectives of the project as it should provide pollen spectra predominantly representative of the wider Lairg area, sufficient to link in with the other sites sampled in the project and with other vegetation histories for Northern Scotland, see,for example, Gear (1989).

There are no inflowing streams and only one small outflow. This makes it particularly suitable taphonomically as there is a reduced element of bias of inwashed pollen from outside the immediate catchment that might otherwise disguise the importance of the rainfall component (Jacobsen and Bradshaw, 1981).

A very full investigation of the lochan basin morphology and sediments was undertaken to establish as far as possible the development of the site. From detailed stratigraphic investigations, it seems that the lochan has probably remained of similar size since it was formed at the end of the last glaciation, no lake sediments being found at or beyond the water's edge. This minimises one taphonomic variable (different sediment types) when interpreting the pollen data.

There are 435 cm of sediment in the centre of the lochan; these consist entirely of highly organic lake muds. Depth of sediment at the loch edge is 709 cm and is composed entirely of peat sediments. Cores were taken from both the centre and the side of the lochan for several reasons. Primarily, it was to establish as fully as possible the pollen stratigraphy taking into account the taphonomic factors previously discussed. By taking two cores, it is hoped that the regional pollen history can be established. The core from the loch edge will show more local bias than the loch centre as it will be greater influenced by vegetation growing on the bog surface. It may therefore be closer to the extra-local definition of Jacobsen and Bradshaw (1981). The lochan centre has been free of these factors throughout development as determined from its gross stratigraphy.

(c). A local site – Achany Glen Site 2 (AG2). National grid reference: NC 576 013

This site is located in the Achany Glen valley, within the *Betula/Quercus* woodland. It is a small woodland peat basin, ca. 30 metres in diameter and therefore fulfils the size criteria for a local site. Analysis of pollen from small hollows within woodland is discussed in detail by Bradshaw (1981). Recruitment of pollen to the site will be greatly influenced by the canopy component and the relative representation of different taxa, due to differential pollen production, is likely to be accentuated at this very local scale (Prentice, 1988).

A second local scale site was chosen for two main reasons. Firstly, to provide a contrast with AG3 which is very closely linked to the archaeological excavation site. The regional site, AG1, differs in scale too greatly to be able to indicate whether the locally based changes identified at AG3, which may be anthropogenic, are more widespread. AG2 should provide this contrast. Secondly, a detailed history of the woodland in Achany Glen should give an important insight into the continuity of woodland cover and ecological change. As the current structure and species composition of the woodland is known and there are historical records of management practices, there is considerable potential for using the site as an analogue for changes in prehistory.

7. CONCLUSION

From the preceding discussion it can be seen that, in applying Jacobsen and Bradshaw's model of pollen recruitment to a site, it has been possible to be objective in the choice of sites selected to fulfil the requirements of the project. In using the model it became clear that their omission of exposure as a criterion could become critical when evaluating some sites. For example, AG1, an extra-local site in model terms, became a regional site when exposure was taken into account. This emphasises the need to allow some consideration of specific factors for the area being studied. The choice of AG2 as a second local site, to complement and contrast with AG3, was based on an assessment of the sites available once the initial survey had been completed and the various research objectives for the project had been considered.

In conclusion, the model has been reasonably successful in selecting sites. Results of the pollen analysis from these sites will be interpreted within the framework defined by the model, enabling issues of pollen taphonomy to be addressed as rigorously as possible.

ACKNOWLEDGMENTS

Many thanks to John Barber and Rod McCullagh for information and hypotheses on the archaeological investigations at Lairg. Thanks also to Dr. Richard Tipping, Professor John Lowe, Margaret Smith and Jean-Luc for reading through drafts of this paper.

BIBLIOGRAPHY

Baillie, M. (1991) Do Irish bog oaks date the Shang Dynasty? *Current Archaeology*. 117, 310–313.

Bell, M. and Walker, M.J.C. (1992) *Late Quaternary Environmental Change*. Longman. New York.

Birks, H.J.B. and Birks, H.H. (1980) *Quaternary Palaeoecology*. Arnold. London.

Bonny, A.P. (1978) The effect of pollen recruitment processes on pollen distribution over the sediment surface of a small lake in Cumbria. *Journal of Ecology*. 66, 385–416.

Bradshaw, R.H.W. (1981) Modern pollen representation factors from woods in south-east England. *Journal of Ecology*. 69, 45–70.

Bridge, M.C., Haggart, B.A. and Lowe, J.J. (1990) The history and palaeoclimatic significance of subfossil remains of Pinus sylvestris in blanket peats from Scotland. *J. Ecology*. 78, 77–99.

Burgess, C. (1990). Volcanoes, catastrophe and the global crisis of the late second Millennium BC. *Current Archaeology*. 117, 325–329.

Dorfler, W. (1992) Radiography of peat profiles: a fast method for detecting human impact on vegetation and soils. *Vegetation History and Archaeobotany*. 1, 93–100.

Dugmore, A.J. (1989) Icelandic volcanic ash in Scotland. *Scottish Geographical Magazine*. 105 (3), pp.108–172.

Gear, A.J. (1989) *The Holocene vegetation history and the palaeoecology of Pinus sylvestris in N.Scotland*. Unpublished PhD.Thesis. University of Durham.

Jacobsen, G.L. and Bradshaw, R.H.W. (1981) The selection of sites for palaeovegetational studies. *Quaternary Research*. 16, 18–96.

Janssen, C.R. (1966) Recent pollen spectra from the deciduous and coniferous deciduous forests of Northeastern Minnesota: A study in pollen dispersal. *Ecology*. 47, 804–825.

Konigsson, Lars-Konig. (1969) Pollen dispersion and the destruction degree. *Bull.Geol.Instn Univ.Upsala N.S.* 1 (6), 161–165.

Lowe, J.J. and Walker, M.J.C. (1984) *Reconstructing Quaternary Environments*. Longman. New York.

Oldfield, F. (1970) Some aspects of scale and complexity in pollen-analytically based palaeoecology. *Pollen et Spores*. 12, 163–172.

Pennington, W. (1979) The origin of pollen in lake sediments: an enclosed lake compared with one receiving inflowing streams. *New Phytologist*. 83. 189–213.

Prentice, C. (1988) Records of vegetation in time and space: the principles of pollen analysis. In: *Vegetation History*. (eds. Huntley, B. and Webb, T. III.), 17–42. Kluwer Academic Publishers.

Tauber, H. (1965) Differential pollen dispersion and the interpretation of pollen diagrams. *Danmarks Geologiske Undersøgelse, Raeckke*. 2 (89), 1–69.

Tauber, H. (1977) Investigations of aerial pollen transport in a forested area. *Dansk Botanisk Arkiv*. 32.

Webb, T., III., Laseski, R.A. and Bernabo, J.C. (1978) Sensing vegetational patterns with pollen data: Choosing the data. *Ecology*. 59, 1151–1163.

APPENDIX 1: CRITERIA AND DESCRIPTIONS TALLIED FOR EACH SITE

A proforma was developed and used for each site in order not to omit attribute data in error.

Date of visit
Map Reference
Grid Number
Distance from Excavation Site
Weather
Aspect
Altitude
Slope
Geology
Nature of Surrounding Topography
Type of Basin
Description of Site
Number of inflowing/outflowing streams
Greatest Width
Greatest Length
Greatest Depth
Presence/nature of infill – broad stratigraphic description of sediment
Description of sediment at base of gouge core
Nature of on–site vegetation
Nature of surrounding vegetation
Practicalities: ie. Accessibility

3. Pollen preservation analysis as a necessity in Holocene palynology

Richard Tipping

1. INTRODUCTION

Assemblages of pollen grains accumulating in sedimentary environments, either natural or archaeological, are predominantly allochthonous. At terrestrial sites, such as peats and soils, a proportion of the pollen can be expected to have originated from within a metre or two of the sampling site, but nevertheless a substantial number of pollen grains are derived from sources at considerable distances from the site.

As such, palynologists have learnt to recognise the complexities in interpretation introduced by, for example, differences in the size of regions depicted in analyses (Jacobson & Bradshaw 1981), differences in the principal routes by which pollen grains arrive at a site (Tauber 1965; Peck 1973; Bonny 1976, 1978; Pennington 1979), differences in the way contrasting sediment types receive pollen (Oldfield, Brown & Thompson 1979; Davis & Ford 1982), and differences in pollen productivity and dispersal (Andersen 1970; Bradshaw 1981).

To these taphonomic considerations, it is necessary to add one further complexity; differences in the state of preservation of pollen within sediments. The problems of pollen recruitment and sedimentation outlined above are essentially geological problems (West 1973), and sub-fossil pollen assemblages are as much to do with geology as botany. Biocoenoses, the totality of organisms inhabiting an environment, are not preserved in sediment. Components are lost or selectively removed, and this transformation of the biocoenosis results in a biased representation, the thanatocoenosis (Brouwer 1978).

One of the most important measures of these biases is preservation state. This has been acknowledged for a great many years now (e.g., Cushing 1964), but despite this, the number of studies which have explored the potential offered by preservation analyses remains disappointingly, and astonishingly, few. This paper will try and illustrate, through the use of selected examples from the author's own work, and others, why the analysis of preservation patterns within sub-fossil pollen assemblages is not only useful, but necessary.

2.THE PRACTICE OF POLLEN PRESERVATION ANALYSIS

Quaternary pollen grains have long been classed as to their preservation state. Cushing (1964) defined a number of types that have, with modifications, formed the basis of most subsequent classifications (e.g., Birks 1970, 1973; Delcourt & Delcourt 1980; Lowe 1982; Tipping 1987). Deteriorated grains can be crumpled, broken or split, corroded (occasionally separated into 'perforation-type' or 'cavitation', and 'exine-thinned' corrosion; Havinga 1964, 1984), and degraded (also called amorphous). Definitions of these are found in, for example, Cushing (1964), Havinga (1964) and Birks (1973). Included in analyses are grains concealed by mineral or organic detritus, though these are not necessarily deteriorated. These classes apply to both determinable grains and grains rendered indeterminable by severe deterioration. Indeterminable grains are not to be equated with unknown grains, which are simply unidentified, but not damaged, pollen types.

The recording of these classes is not without difficulties. Some pollen taxa are highly susceptible to particular preservation classes. For instance, *Juniperus* pollen is often characteristically split, and has been excluded from preservation analysis because of the bias in interpretation this presents (Birks 1973; Tipping 1984). Pollen of the thin-walled Cyperaceae are also more susceptible to crumpling than other grains, and a certain subjectivity is introduced in consideration of what constitutes a crumpled Cyperaceae grain.

More fundamental differences in recording systems are

found, between hierarchical and non-hierarchical classifications. The first ranks preservation classes in a presumed order of significance (normally : degraded, corroded, split and broken; Cushing 1967; Birks 1973; Lowe 1982). The second treats each preservation class as equally significant, and grains are classed on the dominant state of preservation (Delcourt & Delcourt 1980; Tipping 1984, 1987).

The statistical treatment of these preservation classes can differ between analysts, principally in the combining of preservation classes. Most commonly, the major types described above are calculated on a percentage sum which also includes well-preserved grains. But most pollen grains are affected by more than one form of deterioration, and the approach of assigning grains to a single deterioration type, whether hierarchical or not, is an over-simplification. Delcourt & Delcourt (1980) minimised information loss by generating 23 separate classes where preservation classes were combined (e.g., crumpled & corroded; corroded & crumpled), but this methodology leads to sets of data which have limited statistical significance.

A final complication concerns the presentation of data on individual pollen types. The depiction of preservation patterns as percentages of total pollen (or some such general sum) is potentially misleading. Pollen types differ in the susceptibility with which they undergo different types of preservation (Havinga 1964; Sangster & Dale 1964; Konigsson 1969). Changes in the proportions of one preservation class can result from the appearance in the pollen record of a taxon susceptible to that type of deterioration. For example, increases in corrosion in the early Holocene can result simply from the colonisation of a region by *Corylus/Myrica*, a pollen taxon predisposed to corrosion (Tipping 1987).

Despite these complexities, the recording of pollen preservation patterns is comparatively straightforward. The collection of data adds very little time to an analysis, and the examples discussed next will, hopefully, demonstrate that the interpretation of these data can lead to invaluable insights. It can, and should, form a routine and indispensable part of pollen analysis.

3. THE VALUE OF POLLEN PRESERVATION ANALYSES

It is not the object of this paper to present a detailed discussion of the methodology behind pollen preservation analysis. It is considered more important to review what can be gained from the exercise.

The benefits can be broadly divided into two important lines of evidence. Firstly, there is the need to ensure that a sub-fossil pollen assemblage (a thanatocoenosis) is not so distorted and biased by processes of selective loss that it bears no relation to the biocoenosis from which it was derived; that is, that the sub-fossil pollen assemblage is capable of palaeoecological interpretation. This is often an unimportant consideration in 'conventional' polleniferous sediments, such as peats or lake sediments, but becomes of enormous significance in the study of archaeological sediments.

The types of sedimentary environment most commonly analysed in 'conventional' studies have a number of unifying characteristics; they are usually stratified, highly organic, sediment accumulations, deposited in sub-aqueous environments, or environments where the sediment is at least seasonally waterlogged. There are, of course, good reasons for this. Not the least important is that preservation of pollen grains is generally good (although it is argued below that, even in these deposits, preservation analyses are of enormous value).

Archaeo-palynologists are often encouraged to examine contexts that are, with regard to pollen preservation, far from ideal. Highly minerogenic, dry and dusty deposits revealed through excavation are presented for analysis by the archaeologist, in the plaintive hope that sense can be made of the pollen assemblage. However, given the less than perfect milieu, destruction of elements of the original pollen assemblage, even partial, can lead to massive distortions of the pollen record, particularly where, as in most cases, interpretations are based on percentage-derived representations of that assemblage.

It is commonplace to find a textual note referring to the overall state of preservation of the analysed assemblage, but often this appears almost as a sop to criticism, rather than a conscious collation of data. What is almost universally missing from palynological reports is the quantitative analysis of preservation states, and thus an irrefutable demonstration that the assemblage is capable of supporting the interpretations placed upon it.

The second type of analysis employs pollen preservation analysis as a way of gaining information on environmental change. This is less widely appreciated, but examples are presented from lake and peat deposits to show that such analyses contain data that are of real value in palaeoenvironmental interpretation.

The clues to palaeoenvironmental interpretation come from an understanding of the origins of the different deterioration types. However, with the exception of corroded grains, these origins are very poorly understood. This may be one reason for the only limited interest in pollen preservation studies, though it cannot explain why such studies have been neglected in the interpretation of archaeo-palynological contexts.

Deterioration can occur syn-depositionally or post-depositionally. This applies to all types of deterioration. Thus, although corrosion is comparatively well-understood, due largely to the work of Havinga (1964, 1984), in being caused by bacterial attack in the presence of air, corrosion can occur during transport to the site, lying on a soil surface, or when exposed to air due to post-depositional changes in water levels (Tipping 1987). Amorphous pollen grains are commonly thought of as originating in similar ways to corroded grains, but this view probably arose through confusion over these terms in the early literature (Sangster & Dale 1961; Elsik 1966; Brooks & Elsik 1974); there is no clear correlation between corroded and amorphous grains. Some workers (e.g., Lowe 1982) have suggested a relation between the amorphous condition and the longevity of pollen grains within soil profiles, but there is no experimental data on this.

4. PALAEO-ECOLOGICAL INTERPRETATION OF ARCHAEO-PALYNOLOGICAL DATA.

As pointed out above, many archaeological sediments are not conducive to good pollen preservation, and contexts analysed for pollen cannot be assumed in all cases to be capable of palaeoecological interpretation. It is best in such situations to demonstrate, through quantitative data on pollen preservation, that pollen preservation is sufficiently good to allow interpretation.

(1) Buried Soil Analyses

On Biggar Common, an area of unimproved upland pasture near Biggar, in southern Scotland, two recently excavated Bronze Age funerary cairns (Johnston 1998) were found to have been constructed over well-preserved soils. From micromorphological investigation (Tipping *et al* 1994), the soils were seen to be brown earths, with pH of 5.0–5.5; such soils can be seen to be 'marginal' for survival of pollen (Dimbleby 1985).

Fine-resolution pollen analyses (on c. 0.25 cm slices) included routine pollen preservation counts, non-hierarchical in nature. Each identifiable pollen grain was assigned to one of five preservation categories; well preserved, crumpled, broken, corroded and amorphous. In addition, 'light' and 'heavy'deterioration was defined for crumpled and corroded grains; light crumpling was recognised by single folds or creases in the grain, heavy crumpling by folding along more than one axis; lightly corroded grains showed traces of corrosion over less than a quarter of the grain, heavy corrosion over more than a quarter.

Discussion will be confined to the analyses from one of the buried soils, that beneath Cairn III (Tipping *et al* 1994). Figure 3.1 shows the percentage pollen data for the 5 cm of the bAh horizon, and the preservation data for these spectra.

There are a number of indications, from pollen concentration and other measures, as well as from preservation studies (Tipping *et al* 1994), for quite severe distortion of the pollen assemblage, through decay and removal of susceptible pollen types, even within the very limited thickness of soil examined. For instance, the increasing proportions down-profile in Filicales and *Polypodium vulgare*, and reductions in Gramineae and Plantago lanceolata, together with the absence below Sample 7 of a large number of taxa, are thought to be due to differential destruction of pollen taxa. The best preservation is found in Samples 3–7, below which there is progressively greater deterioration, through principally increased breakage and corrosion.

The step-wise decline in Gramineae percentages is matched closely by reductions in well-preserved grains, between samples 4 and 5, and between samples 10 and 11. The representation of other angiosperm taxa is equally explicable by this model. It might be expected that as the numbers of determinable grains are reduced with depth, grains rendered indeterminable would increase. This is not the case. It is tentatively assumed that this is a measure of the severity of deterioration with depth, in that grains indeterminable but recognised as pollen grains in near-surface spectra become not so easily recognised in deeper levels.

Increases in deterioration down-profile do not, however, lead to reduced percentage representation in the two dominant tree taxa, *Alnus* or *Corylus/Myrica*. This is due to the persistence in pollen counts of very intensely deteriorated grains of these taxa, to the point where they are but 'ghosts', but still readily recognisable by their distinctive morphology when not also heavily crumpled. It is the capacity to distinguish these grains irrespective of their preservation state that allows their persistence in the counts. Gramineae grains are not so readily recognisable, and when deterioration reaches levels where no well-preserved Gramineae grains remain (unpublished data), their percentage representation suffers.

From this analysis, and data on other measures, it was concluded (Tipping *et al* 1994) that only Samples 1–7 could be regarded in any way representative of the vegetation prior to monument construction. Samples 8–16 show a large number of indications of severe bias in pollen representation, and cannot be interpreted.

(2) Bronze Age Burial Practices

A number of Bronze Age cists throughout Scotland have now been shown to contain enhanced values of one particular pollen type, *Filipendula*, generally assumed to be *F. ulmaria* (meadowsweet) (Lambert 1964; Dickson 1978; Bohncke 1983; Tipping 1993, 1994). These finds have been interpreted as representing a special deposit within the cist, either a drink, a food or a floral tribute.

The pollen spectra have thus consistently been interpreted as representing a practice contemporaneous with burial rites. However, the pollen samples have most frequently come from the sandy floors of voided cists, and these dry and minerogenic sediments might have led to the differential destruction of susceptible grains, and the enhanced representation of resistant grains; in this regard, *Filipendula* pollen is not noted as being especially susceptible to decay, and its thick wall might indicate survival when other grains are removed.

It is only by routinely analysing the preservation states of the pollen assemblages that this bias can be dismissed. This is not possible for the early analyses, but for the

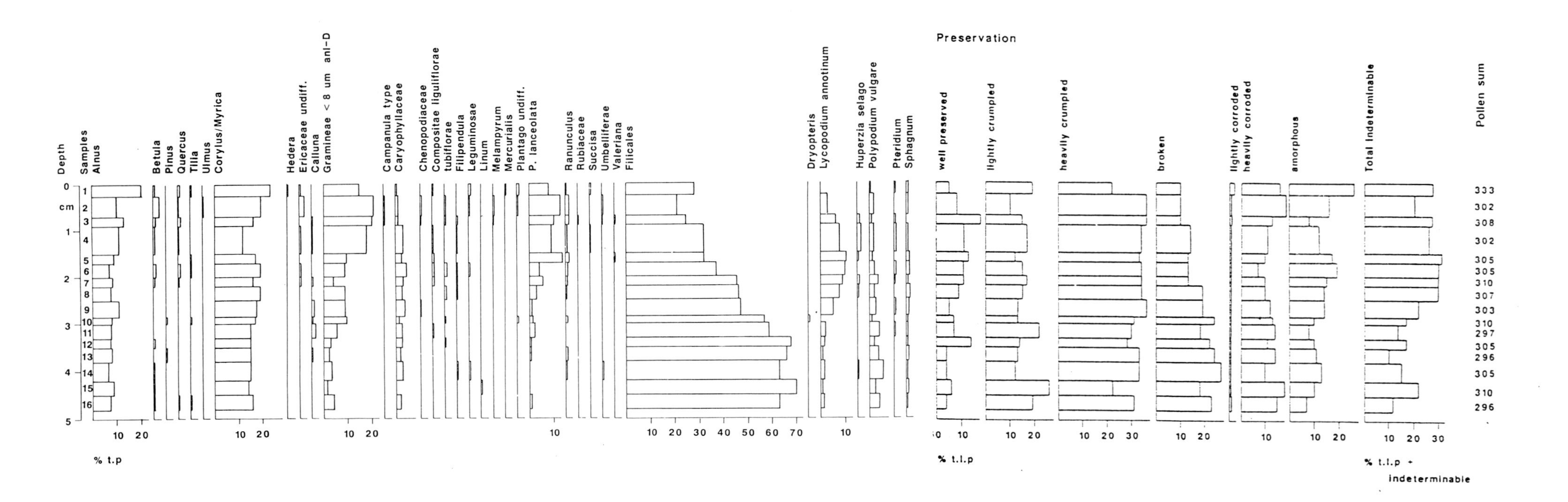

Figure 3.1: Palynological data (pollen percentage data and pollen preservation data) from the buried soil beneath Cairn III, Biggar Common, southern Scotland; Samples 1–16 are c. 0.25 cm sediment slices taken through the 5cm bAh horizon of a buried brown earth.

recently analysed deposits (Tipping 1993, 1994 a) the states of preservation can be readily examined. Table 3.1 shows that, although in all analyses well-preserved pollen grains are a minority of the assemblages, they are a large minority. Physical damage (crumpled and split grains) is the principal form of deterioration. Most importantly, corroded grains are comparatively few. Since corrosion is probably the most significant source of selective removal of grains (Havinga 1964), and can be expected to be pronounced in dry and sandy sediment, such low percentages can be used to demonstrate that high values of *Filipendula* are not a product of differential preservation, and are meaningful.

At a few of these sites, most notably at Beech Hill House, pollen analyses were made of underlying and surrounding sediment sources, as a check on possible contaminants, but these were found to be non-polleniferous. The survival of pollen within the cists, and in relatively good states of preservation, suggests that unusual preservation conditions exist within these cists.

5. POLLEN PRESERVATION ANALYSES AS INDICATORS OF ENVIRONMENTAL CHANGE

(1) Soil instability and the re-deposition of pollen.

Periods of soil instability can lead to the re-deposition of pollen, eroded from soils, in lake sediments. This re-deposited pollen is often more deteriorated than pollen from contemporaneous sources. Re-deposited pollen then has the potential to disguise contemporaneous palynological changes. Many of the early examples of pollen preservation analysis were directed towards distinguishing proportions of re-worked pollen in lacustrine sediments dating to the last glacial period, when solifluction introduced soil material to lake basins (e.g., Cushing 1964; Birks 1970; Lowe & Walker 1977).

Soil instability is a problem in the Holocene also, particularly the later Holocene, when farming communities began to plough soils. Tolonen (1980) in Finland, and Hirons (1983, 1984) in Northern Ireland, used pollen preservation to detect anthropogenic soil erosion and inwashing to lake sediments.

An extreme example of the distortions introduced by re-deposited pollen in lacustrine sediments occurs at Yetholm Loch, at the foot of the Cheviot Hills, in south-east Scotland (Tipping 1992, in press). At this site (Figure 3.2), minerogenic sediment began to enter the lake shortly after the elm decline at c. 160 cm, but at around 2200 cal. BP (138 cm), a very coarse silty sand inwash band was introduced. There is no apparent association of the coarse sediment inwashing with woodland clearance, as might be expected. Instead, the replacement of all major tree types by arable and pastoral herbs, representing an intensively farmed landscape, seems to occur only at the end of the inwashing phase.

The explanation for this anomaly is found in pollen preservation analyses. Figure 3.3 shows the changes in preservation states of one prominent taxon, Gramineae <8 μm anl-D (wild grasses), through the period of coarse sediment inwashing. These curves are typical of patterns prepared for other major taxa, and of changes in total land pollen also (unpublished data). The association of crumpled grains with the coarse sediment band is very clear; consistently greater than 70 % of Gramineae grains are crumpled in this band. Crumpling is a deterioration type to which the balloon-shaped grass pollen grains seem susceptible. The more intense crumpling in this inwashed band need not indicate re-deposition; it could have been induced in contemporaneous grass pollen simply by abrasion with coarse sand particles. However, the peak in corroded grass pollen within the coarsest sediment is likely to originate within dry catchment soils exposed to bacterial attack. Together with other lines of evidence, such as high frequencies of the virtually indestructible spores of Filicales, almost certainly derived from soils (see the Biggar Common example and Pennington (1964)), very large amounts of re-deposited pollen can be detected in the pollen counts. The proportions of these grains seems to have been such that the near-total woodland clearance that did trigger serious soil erosion is not itself registered in the pollen diagram until the eroded sediment supply is exhausted, above c. 1850 cal. BP (125 cm).

Above this depth, well-preserved pollen is increasingly important, and together with sedimentological criteria, can be used to illustrate the absence of later soil inwashing. Above c. 40 cm, the increases in corrosion are closely related to the change from lake mud to peat, and probably represents both syn- and post-depositional exposure of the semi-terrestrial fen to periodically oxygenating conditions.

(2) Climatic Fluctuations

Changes in lake-levels in the Teith Valley, central Scotland, in the early Holocene, have been deduced from increasing proportions of deteriorated pollen, assumed to be derived from basin-edge deposits (Lowe 1982). The patterns of pollen preservation appear similar to those seen in phases of soil inwashing, as at Yetholm Loch, later in the Holocene. Similar fluctuations in groundwater-table, related to changes in effective precipitation, can be detected by pollen preservation changes in peat deposits also. One such site is Burnfoothill Moss, an 8 m deep raised moss from the northern shore of the Solway Firth, near Gretna (Tipping 1995).

Figure 3.4 depicts the changing proportions of well-preserved and deteriorated pollen at Burnfoothill Moss. The figure represents changes in all land pollen grains. Unpublished data demonstrate that the major changes seen in this diagram occur in all the major taxa also, and are not the result of biases introduced by particular susceptible pollen types. Similarities in the patterns of preservation in

Table 3.1: Major pollen taxa (>5% total pollen and spores), and pollen preservation data, for different archaeological contexts at four Bronze Age cists in Scotland; Beech Hill House, Perthshire; Loanleven, Perthshire; Sketewan, Perthshire; Sandfjold, Orkney (see Tipping in press).

	Beech Hill House			Loanleven		Sketewan		Sandfjold	
Context	F78	F79	F80	F7a	F7b	SP1	SP2	C	K
Pollen Taxon									
Alnus	10.8	11.9	8.7	9.7	9.8	5.7	2.7		
Betula	5.5	4.8	3.0	1.0	0.5		0.4	1.5	
Quercus	11.3	4.4	2.3	2.8	3.0	0.5			
Corylus/Myrica	17.0	17.1	8.4	13.9	9.1	17.1	5.3	4.9	
Calluna vulgaris	7.1	5.2	3.4	4.6	1.5	0.5	0.8	8.4	
Gramineae anl-D <8.0 μm	11.7	9.1	10.7	37.4	27.1	20.0	20.1	9.8	4.4
Spergula type	5.2	4.8	5.7						9.5
Bidens type	1.8		0.6	6.7	1.4	0.5	0.4	1.9	4.4
Filipendula	0.6	3.2	5.4	0.2	1.8	0.5	3.8	0.5	4.4
cf. *Filipendula*	17.3	19.1	34.0	4.4	12.0		43.3		40.7
P. lanceolata	1.2	4.8	2.3	4.8	4.1	3.8	5.7	58.1	31.2
Filicales	4.9	4.4	4.0	5.3	11.4	10.0	1.5	1.5	
Polypodium vulgare	1.2		0.3	1.6	1.9	10.5	1.1		
Sphagnum	2.7		0.6	1.9	9.8	22.9	3.0	0.9	1.9
(% total pollen and spores; % t.p.s.)									
Total Pollen and Spores	323	251	297	567	726	210	263	203	157
Total No. Taxa	25	22	24	24	27	20	26	16	12
Pollen Preservation									
Total indeterminable (% t.l.p. + indeterminable)	18.3	19.8	24.8	5.0	6.2	7.0	4.2	22.8	14.6
Determinable : (% t.l.p.)									
well preserved	39.4	22.8	54.6	40.4	38.9	29.4	53.6	25.7	52.1
crumpled/split	47.6	64.9	39.3	46.8	49.3	49.6	39.1	68.1	41.7
corroded	6.6	4.5	2.0	12.6	11.6	17.6	4.0		0.8
amorphous	6.4	8.0	4.1	0.2	0.1	2.5	1.2	6.0	5.3

n.d. - not determined.

different taxa, particularly in taxa which are likely to have occupied very different niches around the pollen site, are probably good evidence for changes to have taken place after the pollen types accumulated on the peat surface, that is, post-depositionally (Tipping 1987).

Burnfoothill Moss has not always been an ombrotrophic moss. Prior to c. 7300 cal. BP (c. 500 cm) the peat was minerotrophic, and supported trees and shrubs. The transition to raised moss coincided with evidence for mesolithic woodland disturbance (Tipping 1995).

This change in peat growth is highly significant in terms of the patterns of pollen preservation exhibited in Figure 3.4. It is seen that corrosion is almost exclusively restricted to the fen-peat. All major taxa below c. 500 cm are corroded to varying degrees. This implies that corrosion occurred after the pollen assemblage, derived from diverse habitats, was deposited on the peat surface (above). The most likely cause of very high levels of corrosion is prolonged lowering of ground-water tables. Fluctuations in corrosion occur, with peaks at the beginning and end of local pollen assemblage zone (l.p.a.z.) BFH B. These are likely to relate to phases of relative aridity on the fen surface, but because much of the decay may be secondary (corrosion occurring at depths below ground surface within aerobic peat), the often abrupt reductions in percentages of corroded pollen, signalling a return to more permanently waterlogged conditions, are more relevant to the dating of presumed climatic shifts.

Virtually no corroded grains are recorded within the raised moss. This is highly significant, since changes in peat and pollen stratigraphies indicate that dry/wet changes on the bog surface continued to occur. Corrosion is thought to set in, particularly with susceptible taxa, in less than a year (Havinga 1964, 1984), and the absence of corroded pollen within the raised moss seems to indicate that periods of relative aridity, although intense, were characterised by large oscillations in groundwater, exposing peat and pollen to air for only short periods. This difference between fen peat and raised moss is probably a function of changes in bog hydrology (Ingram 1982).

Pollen preservation patterns contribute to studies of probable climatic shifts within the raised moss also.

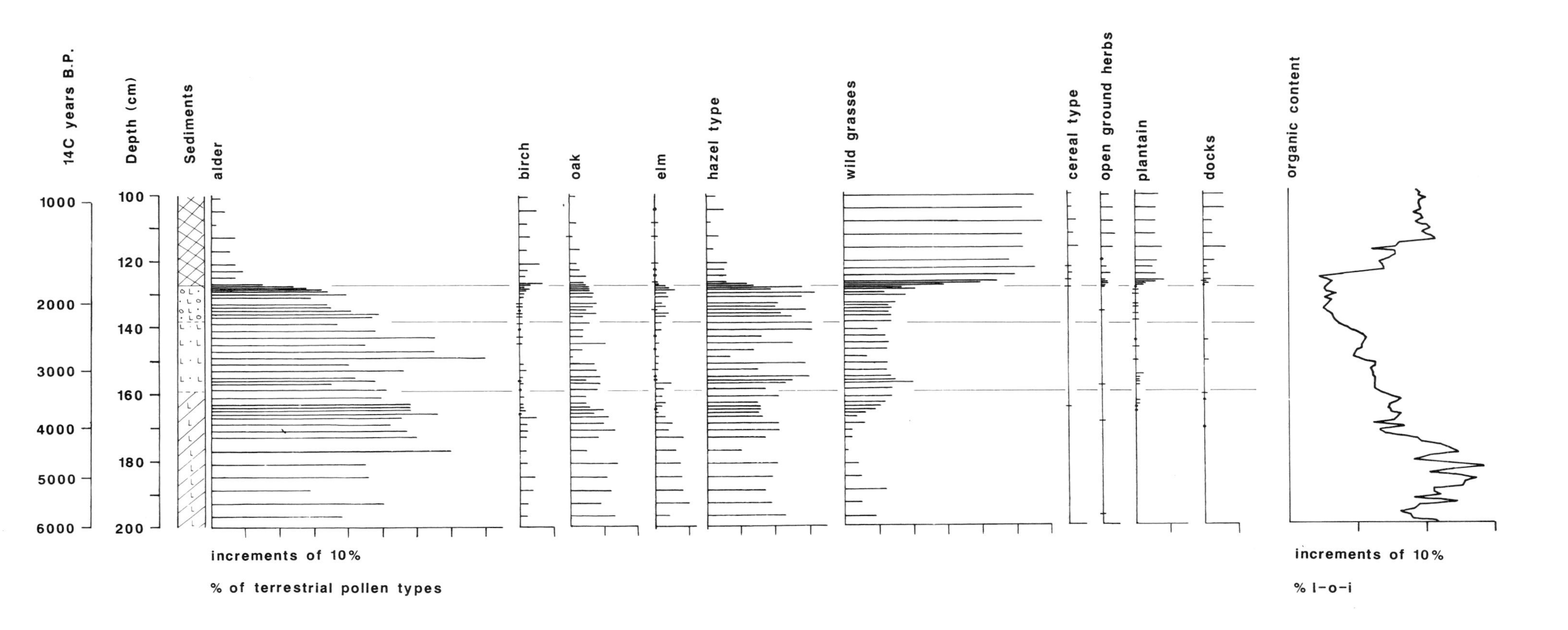

Figure 3.2: Pollen stratigraphy of major tree taxa and anthropogenic indicator herbs in the mid-late Holocene at yetholm Loch, in the Bowmont Valley, Borders Region. Also plotted is the sediment stratigraphy, comprising increasingly clay-rich organic muds (200–160 cm), silty clays (160–138 cm) and clay-rich sands (138–125 cm); highly organic muds replace sands above 125 cm. These changes are also reflected in the organic content curve (see Tipping 1992, in press).

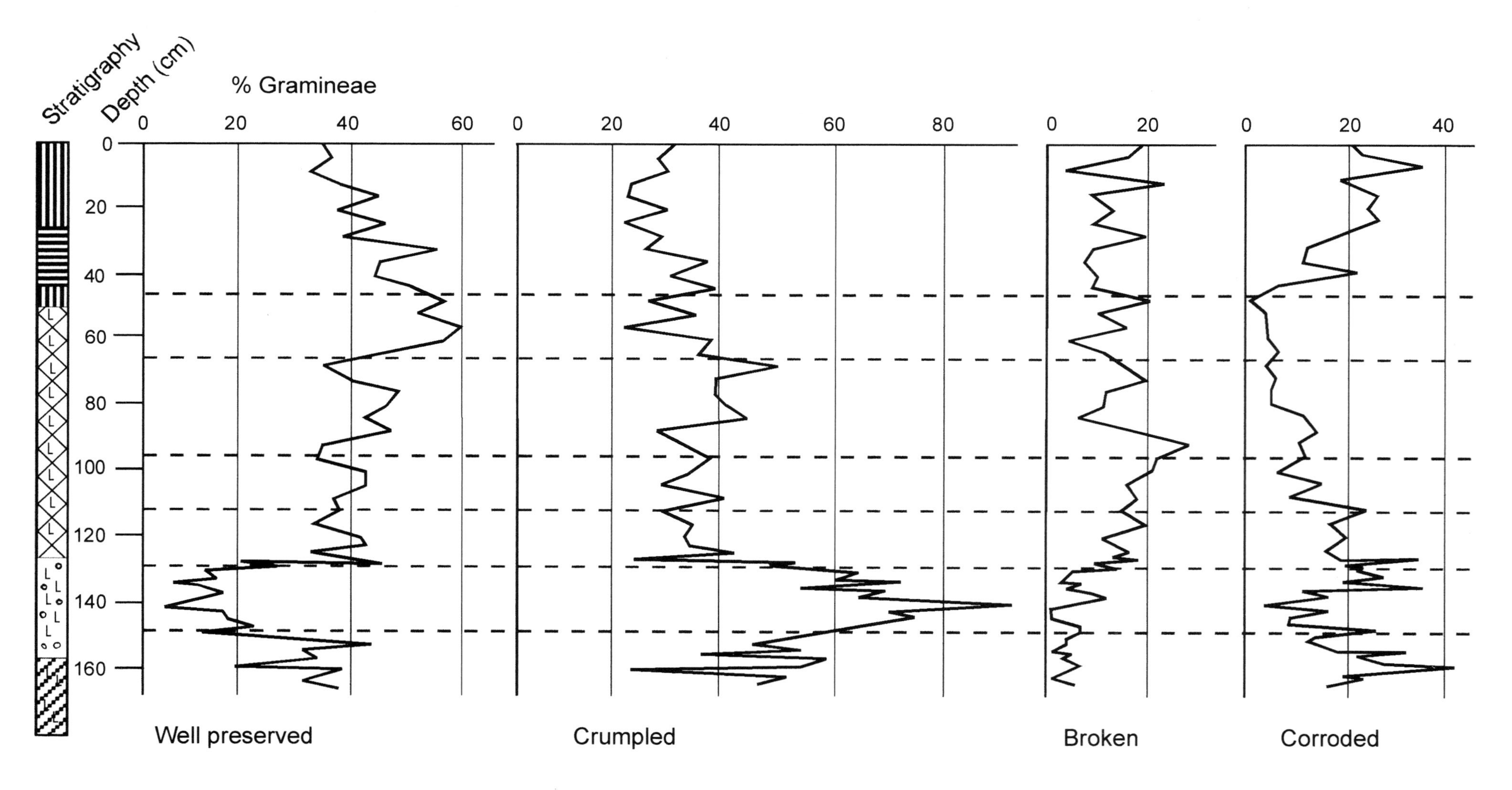

Figure 3.3: Yetholm Loch: pollen preservation patterns for Gramineae <8μm anl-D(wild grasses) between 180–0 cm depth; amorphous grains are too few to be depicted. Sediment stratigraphy is that for Figure 3.2, with peats replacing organic muds above 40 cm.

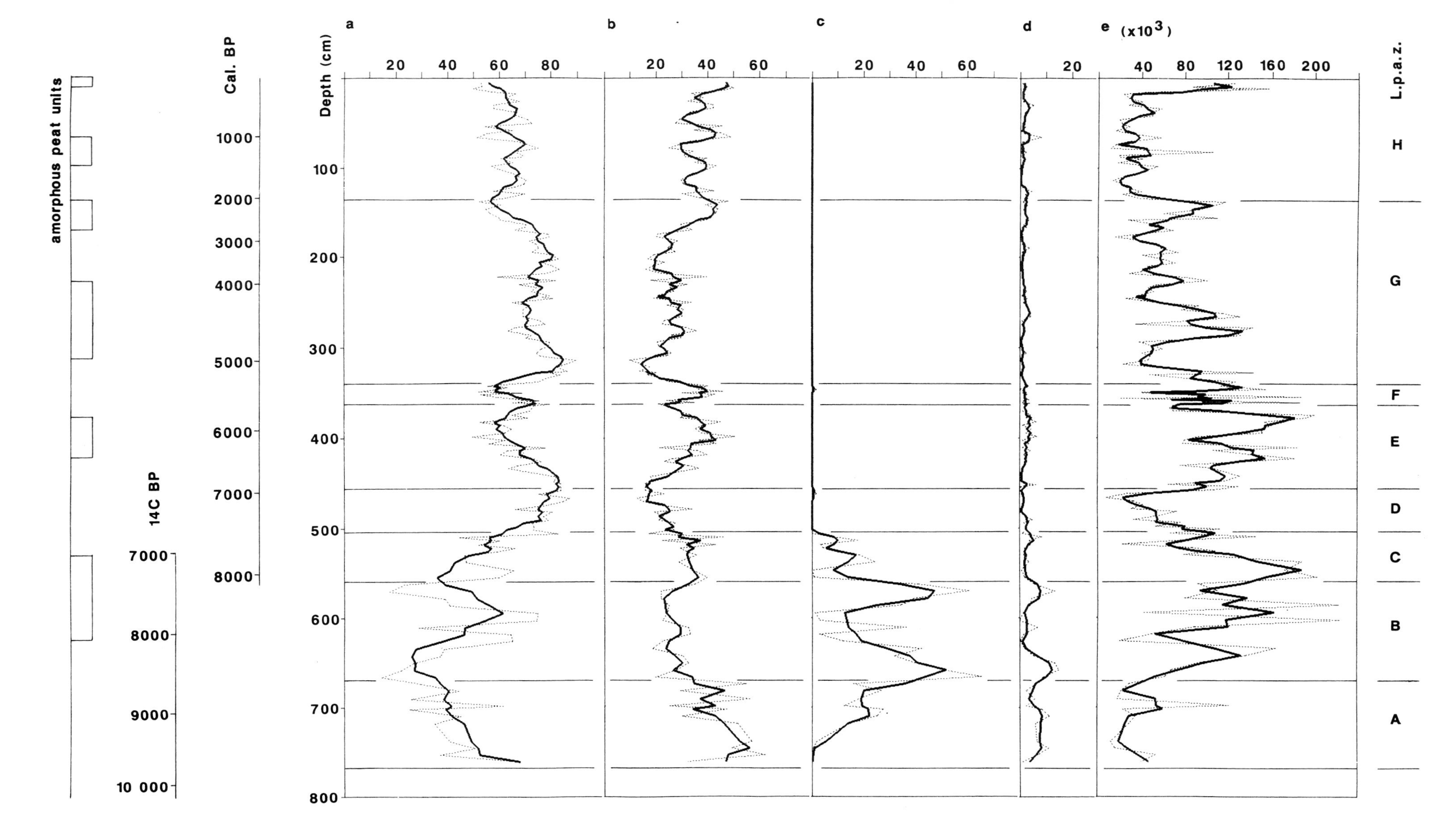

Figure 3.4: Total land pollen preservation and concentration data for Burntfoothill Moss, plotted against years BP (calibrated ages above 560 cm ^{14}C ages below this) and deoth below ground surface: (a) well-preserved land pollen grains; (b) mechanically damaged land pollen grains; (c) corroded land pollen grains, all as % t.l.p.; (d) total indeterminable grains as % t.l.p. + indeterminable grains; (e) t.l.p. concentrations. Curves are plotted as (i) smoothed curves calculated on running means of 3 contiguous spectra (solid lines) and (ii) individual points (dotted lines). Also depicted are local pollen assemblage zone boundaries and sediment-stratigraphic units where amorphous peat is prominent (Tipping 1995).

Preservation is generally excellent, but there are sometimes marked fluctuations in proportions of mechanically damaged grains (Figure 3.4). Unpublished data show that changes in mechanical damage occur virtually synchronously in all taxa >5% t.p., and since these taxa almost certainly grew in different plant communities, post-depositional damage is likely. High proportions of mechanical damage are thus likely to have been induced through sediment compaction (cf. Smith 1984). Compaction in peat is not a direct result of the weight of overburden (Clymo 1984; Middeldorp 1986), and the discrete periods of greater sediment compaction at Burnfoothill Moss are thought to be linked to periods of greater humification, leading to increased pressures on pollen grains within the sediment.

6. CONCLUSIONS

The points raised in this paper are pertinent to any understanding of palynological taphonomy, but they are not new. The examples presented are from recent, but the methodology behind the preservation analyses, and the basis of many of these interpretations, are firmly established.

Yet it is still relevant to raise once again the issue of the need to undertake such analyses. Despite a history of work extending back over 30 years and more, and the oft-expressed recognition of the importance of such work, quantitative data on pollen preservation are still rare. There is probably little reason why this should be the case, given that the principles and working practices of preservation studies can be found in general texts (Birks & Birks 1980; Moore, Webb & Collinson 1992). It is to be hoped that, in future, more attention is paid to what is a fundamental issue in taphonomy and palynological interpretation.

ACKNOWLEDGEMENTS

I would like to thank the organisers of the Durham AEA Conference for inviting me to present this paper, and for their help in publication of the talks. I would also like to thank Professor John Lowe, Royal Holloway College, for triggering my interest in pollen preservation studies.

The work at Biggar Common, Yetholm Loch and the Bronze Age cists was funded by Historic Scotland, and grateful thanks are accorded to Patrick Ashmore for his continuing interest. The analyses at Burnfoothill Moss were funded by the Ann Hill Bequest, through the Dumfries & Galloway Natural History & Antiquarian Society.

Discussions with archaeologists at Biggar Common (Dan Johnston), Beech Hill House (Sylvia Stevenson), Loanleven (Chris Lowe), Sketewan (Roger Mercer) and Sandfjold (John Barber and Coralie Mills) are gratefully acknowledged. In particular, Roger Mercer (RCAHMS) is to be thanked for his inviting me to be involved in his field surveys of the Bowmont Valley (Yetholm Loch) and Kirkpatrick Fleming (Burnfoothill Moss). Stephen Carter worked closely with me in the interpretation of the Biggar Common buried soils. Radiocarbon dates were obtained at the SURRC and NERC facilities at East Kilbride, and the work of Gordon Cook, Douglas Harkness and Brian Miller is much appreciated.

BIBLIOGRAPHY

Andersen, S. Th. (1970) The relative pollen productivity and representation of north European trees, and correction factors for tree pollen spectra. *Danmarks Geologiske Undersogelse* Ser. II, 96, 1–99.

Birks, H.J.B. & Birks, H.H. (1980) *Quaternary Palaeoecology*. London: Edward Arnold.

Birks, H.J.B. (1970) Inwashed pollen spectra at Loch Fada, Isle of Skye. *New Phytologist* 69, 807–820.

Birks, H.J.B. (1973) *Past and Present Vegetation on the Isle of Skye: A Palaeoecological Study*. Cambridge: University Press.

Bohncke, S. (1983) 'The pollen analysis of deposits in a food vessel from the henge monument at North Mains', pp. 39–47 in Barclay, G.J., Sites of the third millennium bc to the first millennium ad at North Mains, Strathallan, Perthshire. *Proceedings of the Society of Antiquaries of Scotland* 112, 39–47.

Bonny, A.P. (1976) Recruitment of pollen to the seston and sediment of some Lake District lakes. *Journal of Ecology* 64, 859–887.

Bonny, A.P. (1978) The effect of pollen recruitment processes on pollen distributions over the sediment surface of a small lake in Cumbria. *Journal of Ecology* 66, 385–416.

Bradshaw, R.H.W. (1981) Modern pollen representation factors for woods in south-east England. *Journal of Ecology* 69, 45–70.

Brooks, J. & Elsik, W.C. (1974) Chemical oxidation (using ozone) of the spore wall of *Lycopodium clavatum*. *Grana* 14, 85–91.

Brouwer, A. (1978) *General Palaeontology*. London: Oliver & Boyd.

Clymo, R.S. (1984) The limits to peat bog growth. *Philosophical Transactions of the Royal Society of London* B 303, 606–654.

Cushing, E.J. (1964) Re-deposited pollen in Late-Wisconsin pollen spectra from east-central Minnesota. *American Journal of Science* 262, 1075–1088.

Cushing, E.J. (1967) Evidence for differential pollen preservation in late Quaternary sediments in Minnesota. *Review of Palaeobotany and Palynology* 4, 87–110.

Davis, M.B. & Ford, M.S. (1982) Sediment focusing in Mirror Lake, New Hampshire. *Limnology & Oceanography* 27, 137–150.

Delcourt, P.A. & Delcourt, H.R. (1980) Pollen preservation and Quaternary environmental history in the southeastern United States. *Palynology* 4, 215–231.

Dickson, J.H. (1978) Bronze Age Mead. *Antiquity* 52, 108–13.

Dimbleby, G.W. (1985) *The Palynology of Archaeological Sites*. London: Academic Press.

Elsik, W.C. (1966) Biologic degradation of fossil pollen grains and spores. *Micropalaeontology* 12, 515–518.

Havinga, A.J. (1964) Investigation into the differential corrosion susceptibility of pollen and spores. *Pollen Spores* 6, 621–635.

Havinga, A.J. (1984) A 20– year experimental investigation into the differential corrosion susceptibility of pollen and spores in various soil types. *Pollen et Spores* 26, 541–558.

Hirons, K.R. (1983) 'Percentage and accumulation rate pollen diagrams from East Co. Tyrone', pp. 95–117. In: Reeves-Smyth, T. & Hamond, F. (eds.) Landscape Archaeology in Ireland.

British Archaeological Reports, 116 *British Series.* Oxford.

Hirons, K.R. (1984) Supplementary palynological data from lakes. *Irish Quaternary Research Association Newsletter* 7, 37–41.

Ingram, H.A.P. (1982) Size and shape in raised mire ecosystems: a geophysical model. *Nature* 297, 300–303.

Jacobson, G.L. Jr. & Bradshaw, R.H.W. (1981) The Selection of Sites for Palaeovegetational Studies. *Quaternary Research* **16**, 80–96.

Johnston, D.A. (1998) Biggar Common, 1987–1993: an early prehistoric funerary and domestic landscape in Clydesdale, South Lanarkshire. *Proceedings of the Society of Antiquaries of Scotland* 127, 185–224

Konigsson, L-K. (1969) Pollen dispersion and the destruction degree. *Bulletin of the Geological Institute of the University of Upsala* New Series 1, 161–165.

Lambert, C.A. (1964) 'The plant remains from Cist 1', pp. 175-178. In: Henshall, A.S. A dagger-grave and other cist burials at Ashgrove, Methilhill, Fife. *Proceedings of the Society of Antiquaries of Scotland* 97, 166–79.

Lowe, J.J. & Walker, M.J.C. (1977) 'The reconstruction of the Lateglacial environment in the Southern and Eastern Grampian Highlands'' pp. 101–118 in Gray, J.M. & Lowe, J.J. (eds.) *Studies in the Scottish Lateglacial Environment.* Oxford: Pergamon Press.

Lowe, J.J. (1982) Three Flandrian pollen profiles from the Teith Valley, Scotland. II. Analyses of deteriorated pollen. *New Phytologist* 90, 371–385.

Middeldorp, A.A. (1986) Functional palaeoecology of the Hahnenmoor raised bog ecosystem – a study of vegetation history, production and decomposition by means of pollen density dating. *Review of Palaeobotany & Palynology* 49, 1–73.

Moore, P.D., Webb, J.A. & Collinson, M.D. (1992) *Pollen Analysis.* Oxford: Blackwell.

Oldfield, F., Brown, A. & Thompson, R. (1979) The effect of microtopography and vegetation on the catchment of airborne particles measured by remanent magnetism. *Quaternary Research* 12, 326–332.

Peck, R. (1973) 'Pollen budget studies in a small Yorkshire catchment', pp. 43–60 in Birks, H.J.B. & West, R.G. (eds.) *Quaternary Plant Ecology.* Oxford: Blackwell.

Pennington, W. (1964) Pollen analyses from the deposits of six upland tarns in the Lake District. *Philosophical Transactions of the Royal Society of London* B 248, 205–44.

Pennington, W. (1979) The origin of pollen in lake sediments: an enclosed lake compared with one receiving inflow streams. *New Phytologist* 84, 171–201.

Sangster, A.D. & Dale, H.M. (1961) A preliminary study of differential pollen grain preservation. *Canadian Journal of Botany* 39, 35–43.

Sangster, A.D. & Dale, H.M. (1964) Pollen grain preservation of underrepresented species in fossil spectra. *Canadian Journal of Botany* 42, 437–439.

Smith, M.V. (1984) Flandrian pollen profiles from the Nar Valley, Norfolk (Great Britain): an analysis of deteriorated pollen percentages using consolidation pressures. *Review of Palaeobotany and Palynology* 41, 283–299.

Tauber, H. (1965) Differential pollen dispersion and the interpretation of pollen diagrams. *Danmarks Geologiske Undersogelse* Ser. II, 89, 1–69.

Tipping, R. (1984) *Late Devensian and Early Flandrian Vegetation History and Deglacial Chronology of Western Argyll.* Unpublished Ph.D. thesis (CNAA), City of London Polytechnic.

Tipping, R. (1987) The origins of corroded pollen grains at five early postglacial pollen sites in western Scotland. *Review of Palaeobotany and Palynology* 53, 151–161.

Tipping, R. (1992) 'The determination of cause in the generation of major prehistoric valley fills in the Cheviot Hills, Anglo-Scottish Border', pp. 111–121 in Needham, S. & Macklin, M.G. (eds.) *Alluvial Archaeology In Britain.* Oxford: Oxbow.

Tipping, R. (1993) 'Palynology', pp. 307–310 in Russell-White, C.J., Lowe, C.E. & McCullagh, R.P.J., Excavations at Three Early Bronze Age Burial Monuments in Scotland. *Proceedings of the Prehistoric Society* 58, 285–323.

Tipping, R. (1994) 'Ritual' floral tributes in the Scottish Bronze Age – palynological evidence. *Journal of Archaeological Science,* 21. 133–139.

Tipping, R. (in press) *Holocene evolution of an upland Scottish landscape : the northern Cheviot Hills.* Society of Antiquaries of Scotland Monograph Series.

Tipping, R. (1995) Holocene evolution of a lowland Scottish landscape : Kirkpatrick Fleming. I. Peat- and pollen-stratigraphic evidence for raised moss development and climatic change. *The Holocene.* 5. 69– 81

Tipping, R., Carter, S. & Johnston, D. (1994) Soil pollen and soil micromorphological analyses of old ground surfaces on Biggar Common, Borders Region, Scotland. *Journal of Archaeological Science.* 21. 387– 401.

Tolonen, M. (1980) Degradation analysis of pollen in sediments of Lake Lamminjarve, S. Finland. *Annales Botanica Fennica* 17, 7–10.

West, R.G. (1973) 'Introduction', pp. 1–3 in Birks, H.J.B. & West, R.G. (eds.) *Quaternary Plant Ecology.* Oxford: Blackwell.

4. Dark earth and obscured stratigraphy

Elizabeth J. Sidell

1. INTRODUCTION

'Dark earth' has been found on many urban sites in Britain. It has often proved enigmatic and difficult to study, although recent applications of micromorphological techniques have provided much information on formation processes (e.g. MacPhail 1981 and MacPhail and Courty 1985). However, the nature of human occupation and activity (if any) taking place during the accumulation of the dark earth is obscured by the difficulty of studying what generally appears as a homogeneous deposit with no visible features or stratigraphy. This has caused a large gap to exist in the understanding of the development and occupation of many urban centres. This paper presents a methodology of studying 'dark earth' and homogeneous deposits in general at a macro scale through the study of the distribution of introduced inclusions. This approach is illustrated by reference to one of the sites (Colchester House) that it has been applied to.

2. THE PROBLEMS

Archaeological stratigraphy in London is very often complicated by truncation of the deposits, which can take many forms. One of the most common occurred during the construction of Medieval and Post-Medieval basements. However, another common situation is in the formation of 'dark earth'. This may obscure or eradicate the latest occupation phases of a site.

Although this paper is not going to enter into the debate upon the nature of what is now commonly termed 'dark earth' (for example see Yule 1990 and MacPhail *op cit*), it is necessary to outline a theory as to its probable origin and the likelihood of truncation and reworking occurring in unconsolidated archaeological sediments. This naturally has a bearing upon the sampling strategy employed in such a case.

It now seems likely that following the disuse of an area, the latest surface may be exposed to weathering and floral and faunal incursion. The resultant activity i.e. animal burrowing, root action etc. will lead to the breakdown of the surfaces which were left exposed. This process will increase exponentially and if allowed to continue undisturbed for some time, an organic-rich deposit ('dark earth') may form although not necessarily requiring that long a period. This will cause a substantial disruption of the uppermost archaeological layers, to all intents and purposes obscuring this latest phasing, often for example, late Roman layers on many urban sites.

The accumulation of the dark earth need not take place under conditions of wholesale abandonment. Material from ephemeral occupation which occurred after the commencement of soil formation could be reworked in the same way as the archaeological layers the dark earth formed upon. If the occupation or activity was of a simple nature, without robust components such as masonry structures, then following this phase, the soil could continue forming and rework any cuts which had been made, i.e. pits, postholes, foundation trenches etc.. These could be totally obscured by mixing caused by biological processes, such as root penetration and animal burrowing, and weathering agents. Such occurrences would slowly blur edges of cuts and gradually eradicate them. Organic material derived from, for example, wattle structures would rot and become incorporated into the organic topsoil. Even stone from rammed gravel paths might be dispersed and become part of the general matrix.

In London and elsewhere, it is this type of occupation which is poorly understood, with few visible traces remaining even when the deposits are well-sealed. Even sites with very robust aspects have been substantially reworked by dark earth, e.g. Union Street in Southwark where a fragment of *in situ opus signinum* was found isolated within a mass of dark earth. However, as the soil develops, and reworking has occurred to a degree, the

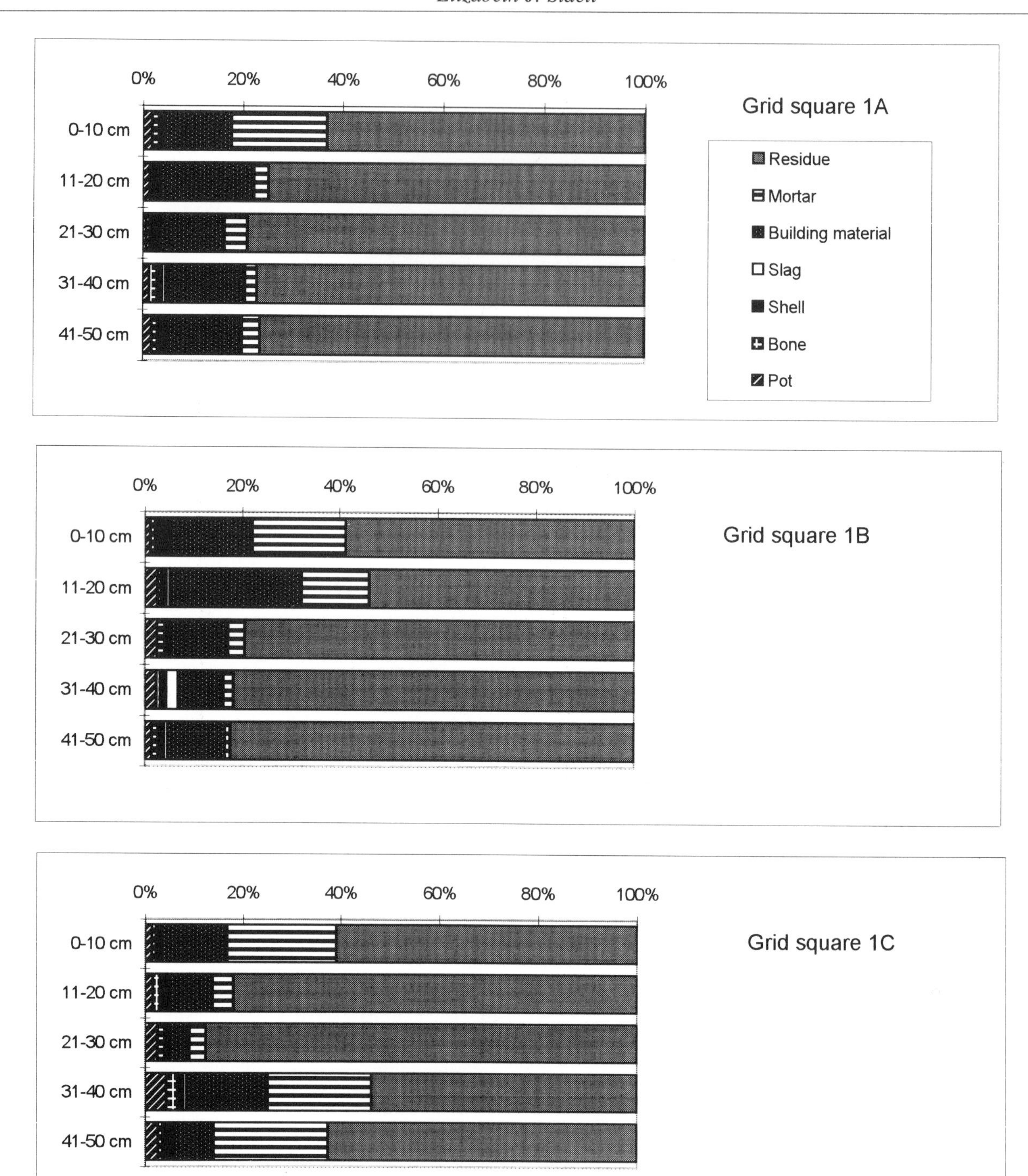

Figure 4.1: Proportions of residue materials from Group 1 samples.

active component of the soil (**A** horizon) will be located further up the profile. This would then seal the earlier material, which will be below the active zone of disturbance. If so, then this should limit the degree of damage caused by the soil-forming processes to the archaeological stratigraphy upon which it was founded and any subsequent activity. Therefore, although features such as pits and postholes may have been obliterated, their contents may have only been dispersed within a limited area. It should thus be possible to isolate this type of pattern by controlled excavation and sampling, and so reveal the phases, and possibly the nature, of occupation.

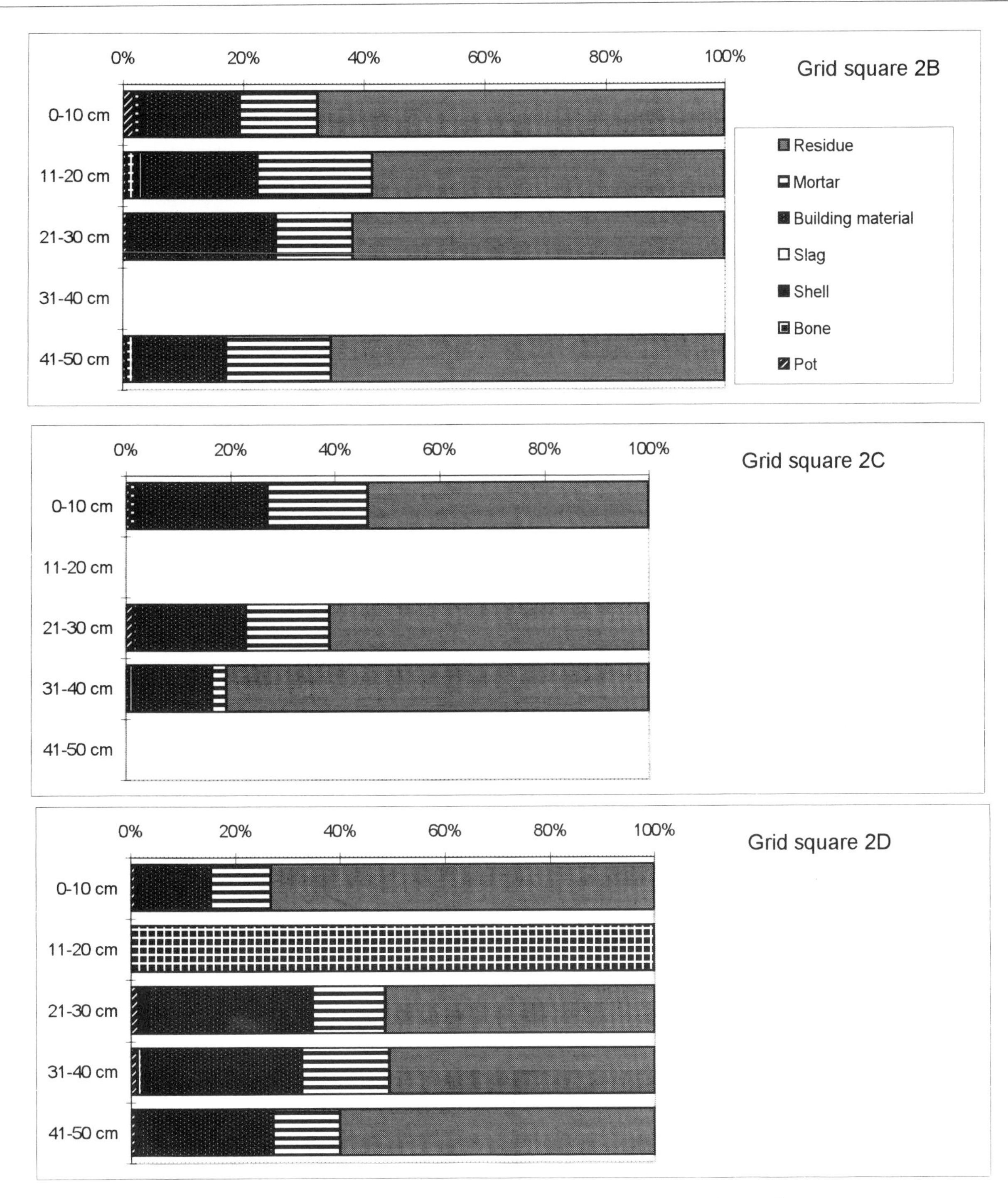

Figure 4.2: Proportions of residue materials from Group 2 samples.

3. DARK EARTH IN LONDON

Sites with dark earth are fairly common in London; recent examples being the Courage Brewery sites, King Edward's Buildings, Guildhall Yard, Colchester House and Bruce House. Therefore the necessity of applying a sampling strategy that was able to locate such activity if present was obvious. It was also necessary that the methodology adopted was feasible within the constraints of working under rescue situations, i.e. juggling excavation areas with the bulldozer driver's timetable. Although micromorphological studies are routinely carried out on such sites, a complementary technique on a larger scale was seen to be desirable. It was decided to adapt the methodology employed by de Rouffignac (1990) at Deansway, Worcester.

It was hoped to outline the strategy that has now been used in London with reference to the two sites where it has been employed. However, due to the vagaries of developer-funded post-excavation projects, it has not been possible to complete the analysis of the material from Bruce House as yet. Therefore it will only be possible to use Colchester House as a point of reference. The site (code PEP89) is located at Tower Hill on the north bank of the Thames. Following a test pit survey in 1989, the site was available for excavation in 1992 in advance of redevelopment and it was possible to dig several trenches on the site. The earliest traces of occupation date to the first and second centuries A.D. Above these early layers, a dark earth developed which was in turn truncated by a massive late Roman building. The remains of this structure include a number of concrete piers, a two metre wide external wall and traces of substantial *opus signinum* flooring, twenty centimetres thick. Sealing this building was a further dark earth deposit sixty centimetres thick. The next phase of activity occurred in the Medieval period when the area was part of the gardens of the Crutched Friars Priory.

The sampling strategy which was implemented isolated a number of squares (50 x 50cm) in the deposit. It was not possible, in terms of both time and money, to apply the technique to the entire area excavated and therefore three distinct locations for the squares were selected. These locations were based on an evaluation of the area in relation to later, major disturbances of the dark earth. One square per area was designated purely for flotation samples, the remainder being used for bulk sieving. The squares for bulk-sieving were excavated in ten centimetre spits which were allocated unique numbers and then 100% sampled. The samples were then passed through 8 and 4mm sieves, the material retained on each was sorted into the various categories (both environmental and artefactual) and individually weighed. The 'residues' from the bulk-sieved samples were retained as a viable category for analysis because it was seen as possibly anthropogenically-derived, rather than an inherent component of the soil and so was a necessary part of the exercise. Possibilities include that the elements of the residue could be a product of rammed floors; in this particular case, the residue seems to be almost entirely derived from the stone debris of the late Roman building. The main categories of material which emerged from the sorting of the samples from this site were pot, bone, ceramic building material (brick, tile, etc.), mortar, shell, slag and residue. These basically divide into two main areas of interest: the mortar, residue and ceramic building material are the product of previous structures, whereas the remainder should be the result of anthropogenic activity. In the case of the Colchester House site, the dark earth contained an extremely large amount of structural debris, and it is therefore possible that some of the fluctuations in the other categories of material have been obscured.

The weights for each category were then converted into proportions of the total weight per sample. It is obviously not possible to compare the weights of the various categories of material within a sample due to the weight difference between the various types of material e.g. ceramic building material and oyster shell. Therefore the pattern found for all the components of a sample have been compared across spits and squares. This sort of clustering is similar to that used in field-walking exercises: two-dimensional recording of finds to locate sites. The dark earth sampling strategy as employed now in London uses three-dimensional recording of material in the attempt to locate hitherto obscured features by looking at fluctuations in sample composition.

4. RESULTS

Three groups of squares were used on the Colchester House site. Two were located within the main trench and have been designated 1 and 2. The third group (3) was in a second trench, approximately twenty-five metres to the east. It was decided to experiment with the positioning of the squares in relation to each other, and so the squares of one group were separated by fifty centimetres, and in the remaining groups the squares were directly adjacent to one another.

Group 1

This group comprised four squares (a,b,c and d); the first three for bulk-sieving and the last for flotation.

They were located at the north end of the main trench, each square separated by fifty centimetres. Generally the results for the three bulk-sieved squares were similar, with the pot, bone, slag and shell remaining in fairly consistent proportions throughout the spits (Figure 4.1). The upper levels of all three show a high proportion of mortar and ceramic building material, possibly the traces remaining of later robbing. Square c also showed high values of these categories in its lowest two levels, although this material was generally absent in the lower levels of the other two squares. These just show a predominance of residue. It is a possible that what is observed in square c are the battered remains of one of the concrete and mortar piers, a wall, tile floor or something of this nature. Otherwise, there do not appear to be any other major patterns in this group.

Group 2

This group was located at the south end of the main trench, approximately fifteen metres from group 1. Again four squares were gridded out (a-d), with 2a designated for flotation samples. The squares were all placed directly adjacent to each other in a north-south line. Unfortunately, there has been some data loss in this group, and so it was impossible to complete the charts. However, some examination of the results is possible (Figure 4.2).

As with group 1, there is very little fluctuation in the bone, pot, slag and shell proportions over the area, within

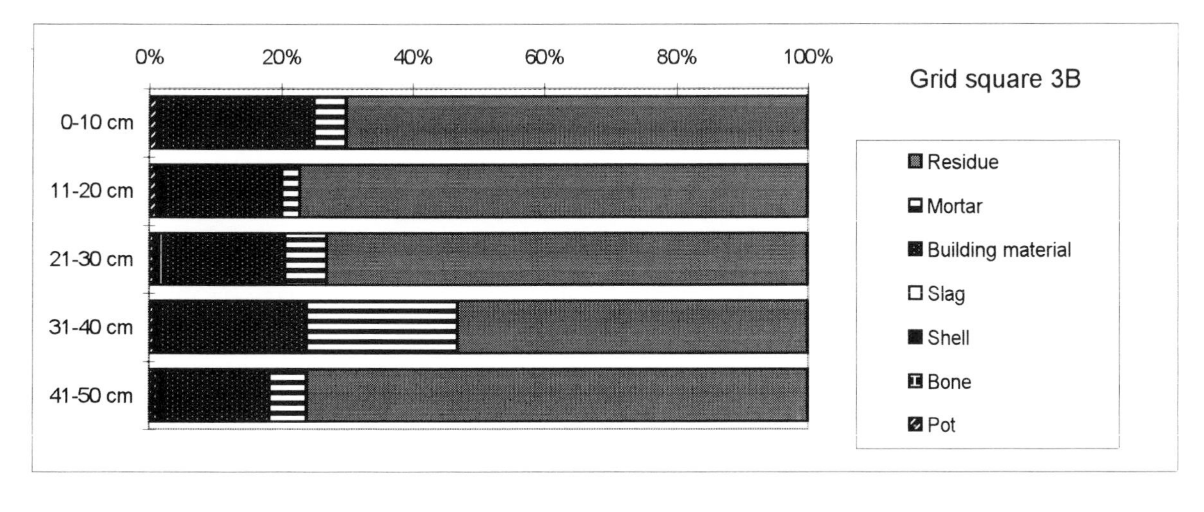

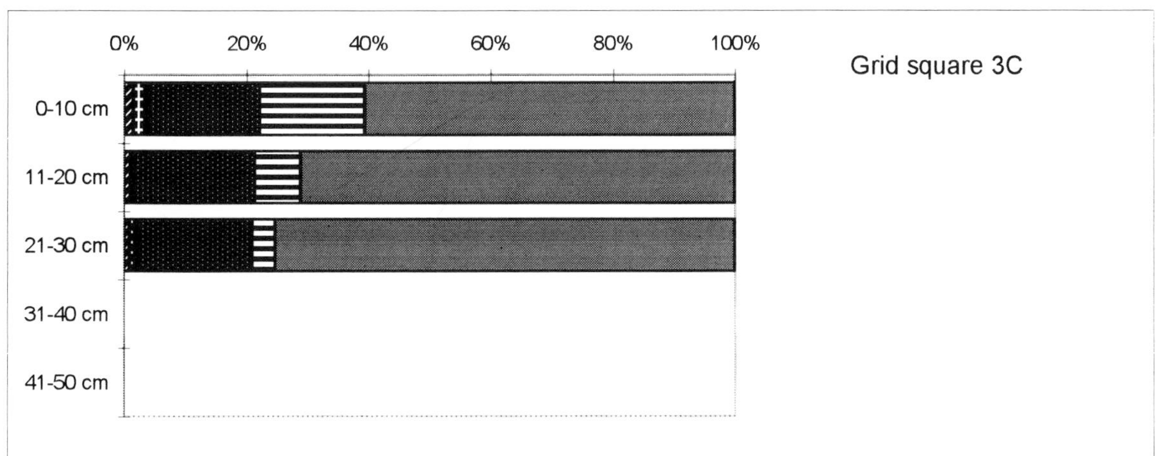

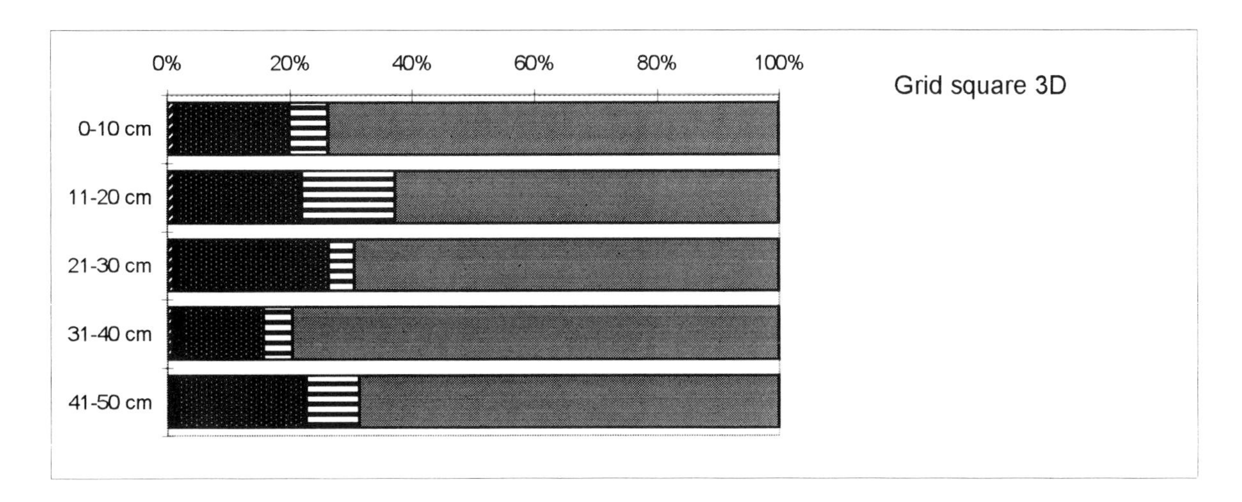

Figure 4.3a: Proportions of residue materials from Group 3 samples.

the profile. However, the proportions of mortar, ceramic building material and residue are so great throughout the sequence that it is possible that the issue has been clouded. What is apparent though is a similarity between squares d and b, with the exception of level 11–20cm in 2d which is more or less a pure bone deposit. Proportions of mortar and ceramic building material increase as one moves down the profile, with a consequent decrease in the proportion of residue. This may be indicative of robber activity in the middle of the accumulation of the dark earth. Although there is some information missing from square c, it does not immediately follow this pattern. There is a marked increase of mortar, and a slightly less dramatic increase in ceramic building material at the expense of residue moving

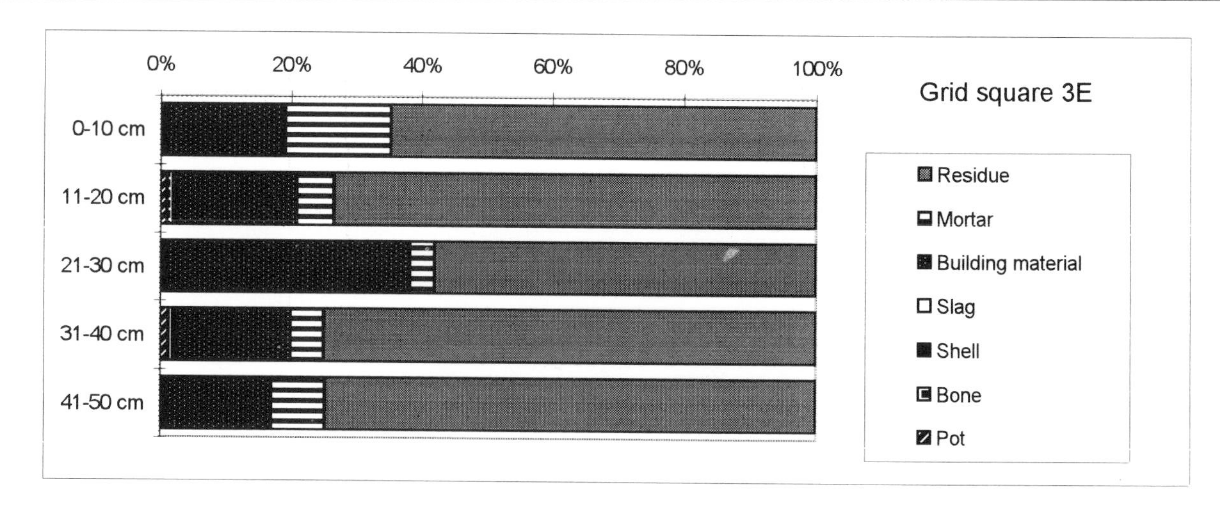

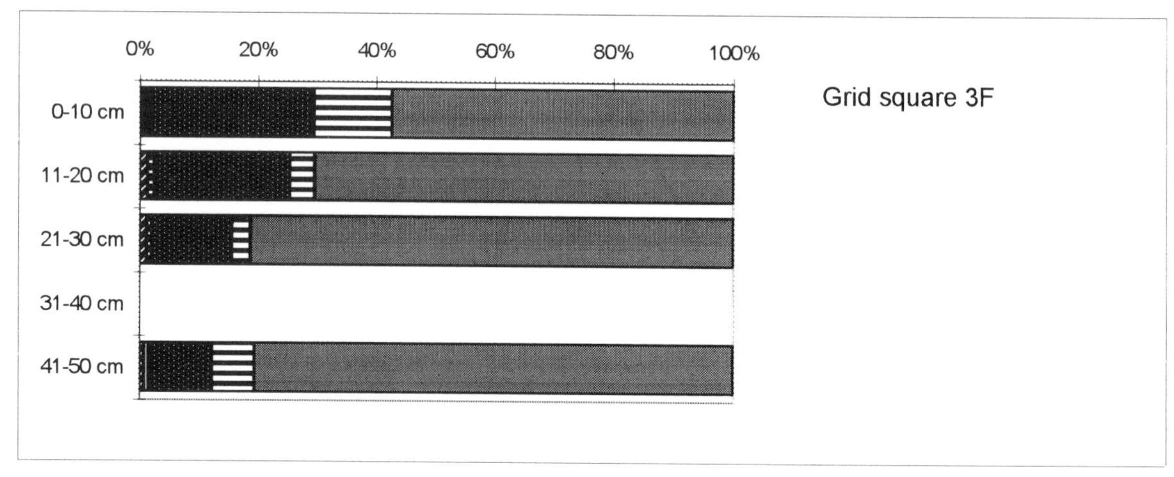

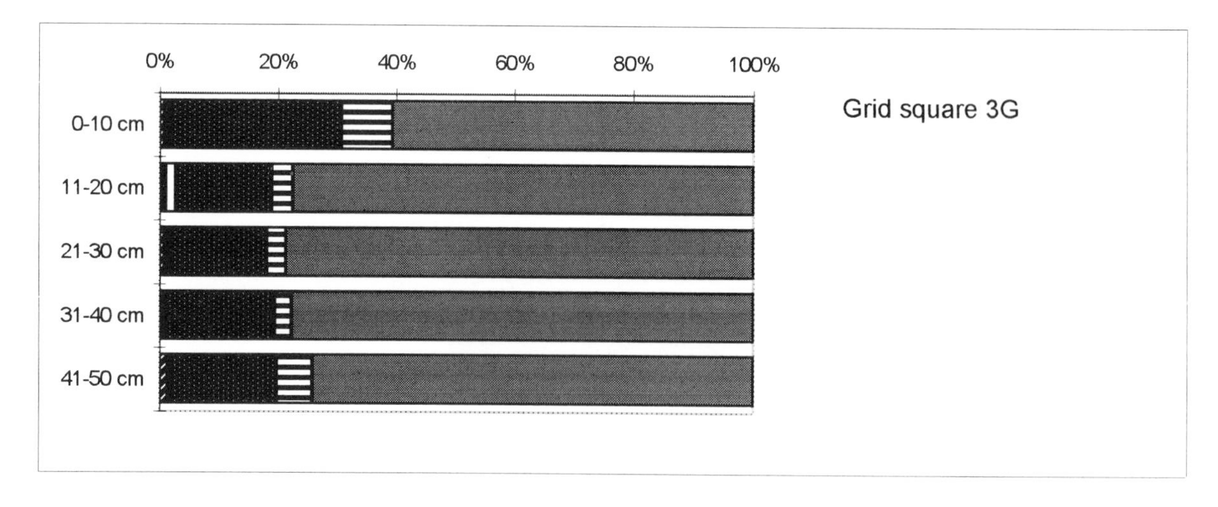

Figure 4.3b: Proportions of residue materials from Group 3 samples.

up through the profile. Again this may represent a late robbing of the later Roman building. However, it is also possible that the material had been dumped here. Obviously dating evidence of the ceramics will be essential to aid interpretation in cases like this.

Group 3

This group was located in the smaller trench, approximately twenty-five metres to the east of group 2. Eleven squares were laid out (a-k), two of which (a and i) were only sampled for flotation. Again there appears to have

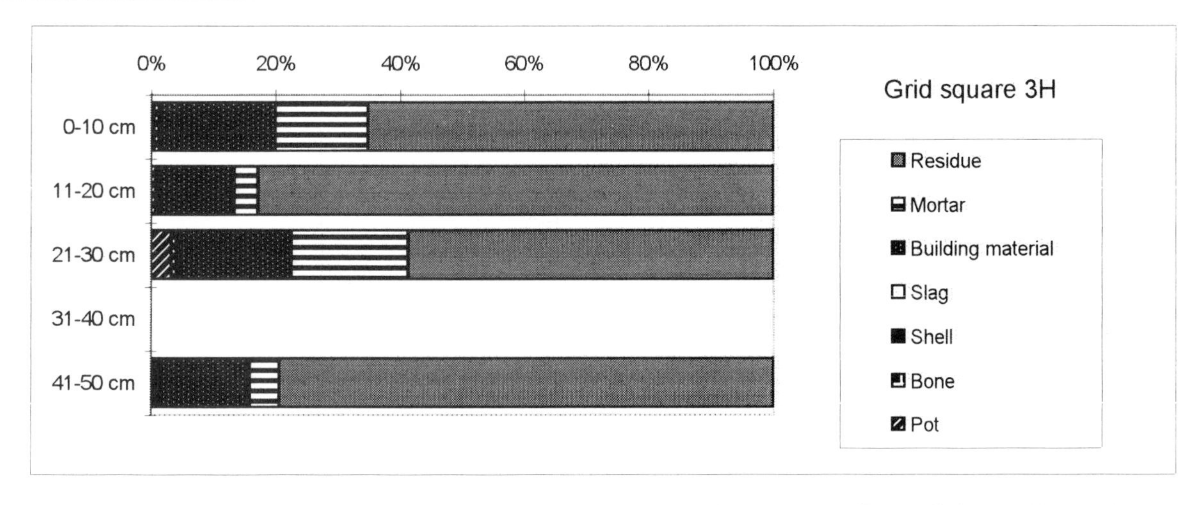

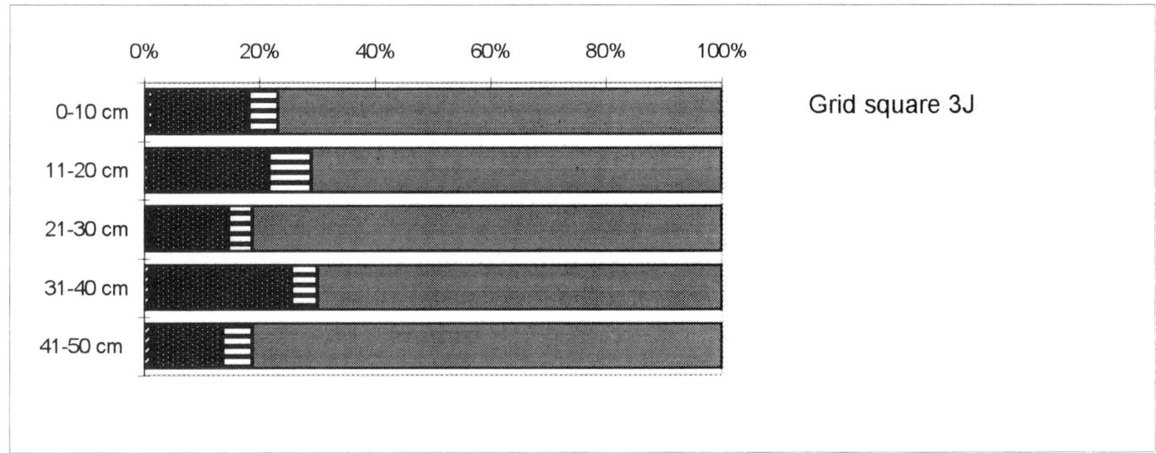

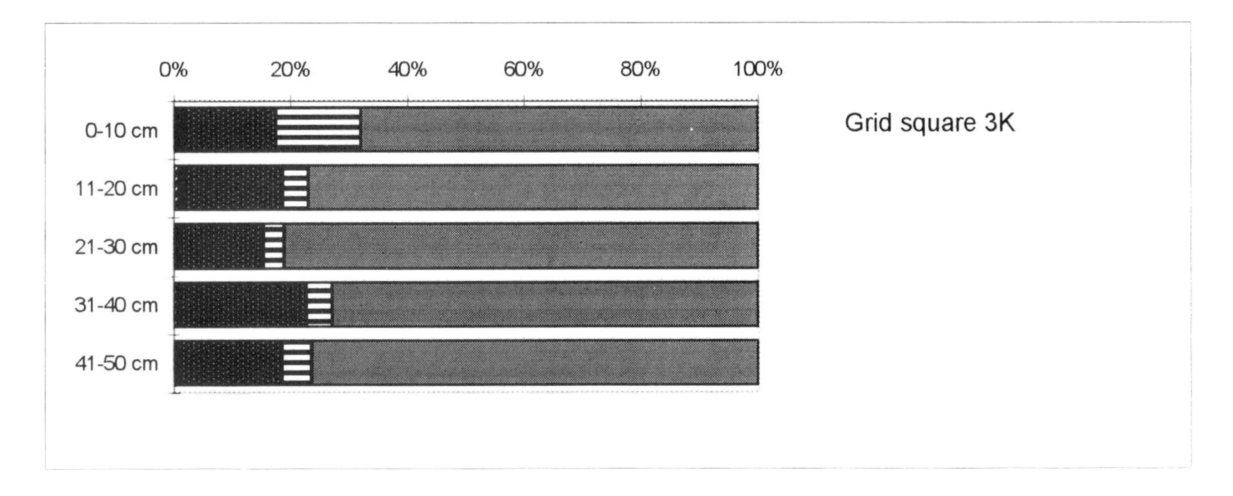

Figure 4.3c: Proportions of residue materials from Group 3 samples.

been the odd sample mislaid, but not to such an extent as in group 2. Squares a and b were slightly isolated from the rest by approximately three metres. Squares c to k were all located directly adjacent to each other in a block of three by three, with square i taken for flotation.

Grid square b was a bulk-sieved square with a fairly consistent composition pattern through the profile (Figure 4.3). A slight fluctuation occurs in the fourth spit where there is an increase in mortar. Again this may represent stone robbing from the late Roman building, leaving the mortar fragments behind. The remaining eight squares show the same consistency of shell and slag (Figure 4.3). However, there is more fluctuation in the proportions of bone and pot than has been demonstrated previously. This may indicate more activity on the site in the form of deposition of domestic waste. One point to note is that the details of the identification and analysis of the various groups of material needs to be introduced into the discussion of this type of exercise. Otherwise, a proportion fluctuation could be caused by an 'accident' i.e. a dead animal, rather than by humanly-derived deposits such as food waste. This could lead to all kinds of mis-interpretations and should therefore be prevented by a full integration of all available information incorporating data from the individual specialists.

Identifying specific occurrences is quite difficult in this block. Squares d, e, h, j and k are all fairly enigmatic with fluctuations in ceramic building material and residue throughout the profile, whilst the proportion of mortar remains generally constant. This is basically a splitting of the various structural components through the deposit. Without a fluctuation in the proportion of mortar, it is not really possible to attribute the others to robbing. However, there may be traces of robbing in squares c and f where there is a marked increase in mortar up the profile. The composition of the spits in square g is extremely consistent except in the uppermost layer where there is an increase of ceramic building material. This may be a later dump. It can be seen that there is a split across this block, with the western edge showing an increase of mortar and ceramic, possibly indicating later robbing, whilst the eastern side shows peculiar fluctuations which are more difficult to interpret.

5. CONCLUSIONS

The initial results of this exercise have demonstrated the value of studying the non-edaphic components of dark earth deposits. They have the potential ability to isolate occurrences which would probably have been overlooked using conventional excavation techniques. Such a taphonomic study is obviously of great use from the standpoint of understanding the environmental and economic development, but is also of great importance as an integrated, i.e. stratigraphic and artefactual, approach to understanding the development of the site as a whole.

Several points have arisen from the analysis of the results from Colchester House, notably the dominance of the residue. This may be a particular factor of this site incorporating the demolition debris from the late Roman building. It is therefore suggested that two sets of calculations are used in future - one as here and a second one excluding the residue category from the total weights. In addition, the need to have specialist input concerning the specific nature of the categories of material is essential to avoid the possibility of misinterpretation.

Obviously the results of this system need to be analysed from another site before commenting on the degree of success. This will occur on the material from the Bruce House project, which took place on a much larger scale. Twenty square metres located adjacently were treated with this sampling strategy, and so with the existing knowledge of the idiosyncrasies of the method, it should be possible to demonstrate the full potential. Clearly a pilot study is needed, with clear results to justify the additional expense incurred by developers for this type of exercise. However, these initial results are promising, and great potential exists for increasing our knowledge concerning the occupation and activities which may have taken place during the formation of dark earth.

ACKNOWLEDGEMENTS

Thanks must be extended to all those who participated in this project, notably James Rackham and Dave Sankey, and also Ian Tyers and Keith Wilkinson for their combined computer expertise.

BIBLIOGRAPHY

McPhail, R.I. 1981 Soil and botanical studies of the dark earth, pp. 309–31. In: Jones, M. and Dimbleby, G. (eds) *The Environment of man: the Iron Age to Anglo-Saxon period.* British Archaeological Reports, British Series 87. Oxford.

MacPhail, R.I. and Courty, M-A. 1985 *Interpretation and Significance of Urban Deposits* pp. 71–84. In Edgren, T. and Jungner, H. (eds.). Proceedings of the third Nordic Conference on the Application of Scientific methods in Archaeology. *ISKOS* 5. Helsinki.

de Rouffignac, C. 1990 Analysis of 'Dark Earth' from the Bull Entry site. Deansway Archaeological Project. *Deansway Interim Research Paper 1.* Hereford and Worcester County Museum.

Sankey, D. 1993 Excavations at Colchester House Assessment. Museum of London Archaeology Service archive report. London.

Yule, B. 1990 The 'dark earth' and Roman London. *Antiquity* 64, 620–28.

5. Through a taphonomic glass, darkly: the importance of cereal cultivation in prehistoric Britain.

Peter Rowley-Conwy

1. INTRODUCTION

This contribution will address two related questions:

(1) Why is the evidence for prehistoric cereal cultivation in Britain generally less spectacular than the evidence from mainland Europe?

(2) How good is the evidence that cereal cultivation in the neolithic and early bronze age of Britain was relatively less important than it was in later periods?

It will be argued that many of the observed differences between the areas and periods may result from (a) variable frequency of charring, connected with storage and processing practices, and site function; and (b) variable efficiency of recovery; in other words, that taphonomic factors may be magnifying or wholly creating a dichotomy which did not exist in the past.

2. CEREAL CULTIVATION IN BRITAIN AND EUROPE

(a) The Continental Comparison

Spectacular finds of charred cereal remains have long been a major aspect of the archaeological record of mainland Europe. Denmark was an area where much early work took place, thanks to the efforts of archaeologists such as Gudmund Hatt, and the botanists Knud Jessen and Hans Helbæk. By the middle years of this century, a series of major cereal finds had been excavated. These came from sites such as Alrum (Hatt 1941), Dalshøj (Klindt-Jensen 1951), Fredsø (Hatt 1931), Ginderup (Jessen 1933, Hatt 1937), Nørre Fjand (Hatt 1941), Østerbølle (Helbæk 1938) and Sorte Muld (Klindt-Jensen 1951). What is impressive is not just the large number of sites – Denmark was a country which pioneered archaeobotanical research at an early date - but the sheer quantity of material that the sites provided. Even a reading of Helbæk's (1954a) short review reveals several mentions of cereal finds measuring several tens of litres.

The archaeological record in Britain was dissimilar: when Helbæk examined the British material he listed a number of charred grain finds (Helbæk 1952a, 229) among which only those from Itford Hill and Fifield Bavant were of even moderate size. Much of Helbæk's British study (op. cit.) dealt with plant impressions in pottery; here the British data were fully comparable to the Danish, and were used for comparative purposes by economically inclined prehistorians (e.g. Clark 1952, 108). With regard to spectacular charred finds, however, Britain was not in the same league.

Remarkably, this situation continues today. "Remarkably" because flotation has been used relatively widely in Britain, at least in comparison with Denmark and our other mainland neighbours. Despite this, major finds continue to turn up on the mainland (some Danish instances will be discussed below), while in Britain the "big finds" remain elusive and rare. Typically, British cereal samples are smaller and more diffuse, often containing a range of other items such as weed seeds and sometimes chaff fragments. Such samples are a characteristic product of flotation, because some flotation systems can process large quantities of excavated material (e.g. that described by Jarman, Legge and Charles 1972). For example, the writer's experience with a froth flotation unit (an improved version of that described by Jarman *et al. (op. cit.)* designed by A.J. Legge) is that up to 50 buckets of spoil from a single large context can be processed per hour. At 10 litres per bucket, this is 500 litres per hour. If plant remains are present at a density of say one item per litre of excavated material, then a large sample of low-density material may be recovered between breakfast and the morning tea break – without the excavators in the trench being aware that they have found any plant remains at all. (This, one may add, emphasises how important it is that flotation of *large* samples should be carried out on all excavations.)

(b) Low Density Samples in Denmark

In sum, therefore, in Britain archaeobotanists appear to work harder but for less visible and spectacular results. This visible dichotomy is however only partially a reflection of the real situation: Britain may lack the spectacular continental samples, but on the continent the diffuse and heterogeneous samples are there if you look for them. This section discusses samples the writer was fortunate to work on from Denmark.

Three bronze age sites are considered. Voldtofte is a large settlement dating to the late bronze age (Berglund 1982); the writer recovered a sample of botanical material using froth flotation from a large and rather undifferentiated rubbish midden (Rowley-Conwy 1982). Egehøj comprises three longhouses of early bronze age date (Boas 1983); the excavator observed cereals during excavation and samples of spoil were bagged and stored, later to be put through the writer's froth flotation unit (Rowley-Conwy 1984). Finally, Lindebjerg consists of a single house site, also of early bronze age date (Jæger and Laursen 1983); the excavator, Mr. Anders Jæger, recovered a very large quantity of cereals by hand during the excavation, and the writer later examined this and samples recovered by froth flotation (Rowley-Conwy 1978).

The three assemblages are summarised in Figure 5.1. Egehøj and Lindebjerg are both major finds similar to the earlier ones described in the previous section, and were observed during excavation. Both sites yielded nearly pure cereal finds with hardly any weeds. At Voldtofte, however, a much wider range of species was recovered, including many conventionally regarded as weeds. Botanical items were not visible during excavation, but were all recovered by flotation of the midden deposits. The quantity of material put through the froth flotation unit was recorded, so the density of seeds can also be calculated for the Voldtofte and Egehøj samples. At Voldtofte, density was (except for *Chenopodium album*) uniformly low, never approaching the concentrations found at Egehøj (Figure 5.2).

Voldtofte is therefore a classic flotation sample, heterogeneous and of low density, which would not have been recovered by conventional Danish means. Nor is it unique in Denmark. Figure 5.3 plots two samples of Viking age date. The Fyrkat sample came from inside a longhouse in the Viking fortress. It comprised some 70,000 grains of rye with hardly any admixtures, and was recovered during conventional excavation (Helbæk 1970, 1974). The Øster Aalum sample was, unusually for a sample of this type, also recovered during excavation by an alert excavator, Mr. David Liversage. The sample is interpreted as the out-sweepings from a hearth or hearths, fortuitously preserved at a relatively high density because it was dumped in a pit (Rowley-Conwy 1988).

Voldtofte and Øster Aalum thus demonstrate that low density, heterogeneous samples are to be found in Denmark. It would indeed be highly surprising were this not the case: following ethnographic work by Hillman (1981, 1984) and G. Jones (1984), such samples are interpreted as the waste products of crop processing activities. All agricultural communities would have generated such waste, which would from time to time have become charred. As a result, it is concluded that part of the perceived difference between Britain and Denmark can be accounted for by recovery method: in Britain flotation (albeit usually of small samples) is used rather more frequently than it is in Denmark, as a direct result of which the low density, heterogeneous samples are found more often; while in Denmark recovery methods are usually inadequate to recover anything but the largest, most obvious samples.

(c) Why so few Big Finds in Britain?

Recovery cannot be blamed for the rarity of large, high-density finds in Britain; some other factor must be sought. Context must be considered. It is striking how often the large, high-density finds come from burnt out houses. Among those mentioned above, this was the case for the Viking sample from Fyrkat; for the bronze age samples examined by the writer from Egehøj and Lindebjerg; and for the earlier finds of Alrum, Ginderup, Østerbølle, Nørre Fjand, Dalshøj, Sorte Muld and Fredsø. In short, every large Danish find considered here comes from such a context (see above for references).

It is unlikely that such large finds represent material being processed or cooked when the houses were burnt - the samples are too large, and are often spread over large areas of the floor. It is more likely that the samples come from material stored in the houses and destroyed with them. Some support for this comes from the fact that different cereal species sometimes predominate in different samples within the same house (e.g. Egehøj: Rowley-Conwy 1984, fig. 5). This suggests that the various cereals were not intermingled, so that some of the spatial separation of the storage survived the destruction of the building.

It is unlikely that major cereal storage would take place at ground level. Quite apart from getting in the way during day-to-day activities, damp would be a problem. At Egehøj, the major cereal concentration was found at the east end of the longhouse, where the floor had been dug down some 30–40 cm below the rest of the floor level. This area would probably be more damp than the rest, thus exacerbating the problems of storing cereals at ground level. It is more likely that cereals were stored in bulk in the roof space. Heat rising from hearths on the floor would rise to fill this space, creating a warm and dry atmosphere suitable for storing cereals. Smoke from these hearths would permeate the roof space and might discourage insect infestation. In this scenario, accidental burning of the house would result in charred cereals arriving at ground level where the excavators find them only during the final collapse of the structure.

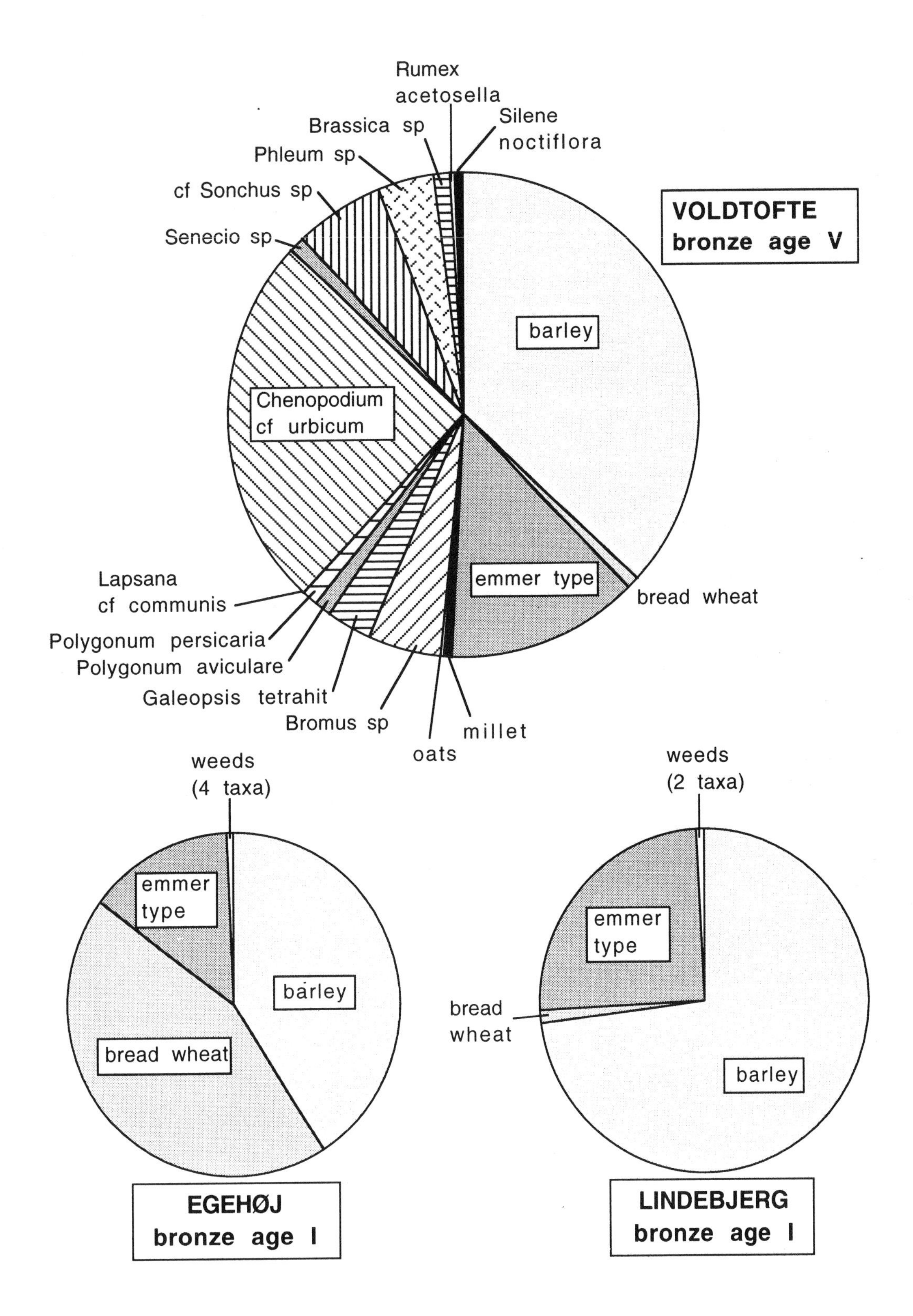

Figure 5.1: Contents of botanical samples from three Danish bronze age sites. Voldtofte from Rowley-Conwy 91982, fig. 1); only the main sample is used, and the coarse and fine fractions are multiplied up and combined. Chenopodium album is the most common plant (see Figure 5.2) but is not included here as its large numbers would compress most of the other plants to the point of invisisbility. Lindebjerg from Rowley-Conwy (1978, fig. 1); only the sample from the structure is used. Egehøj from Rowley-Conwy (1984, table 1), all samples amalgamated. Unidentified wheat and unidentified cereals are not included.

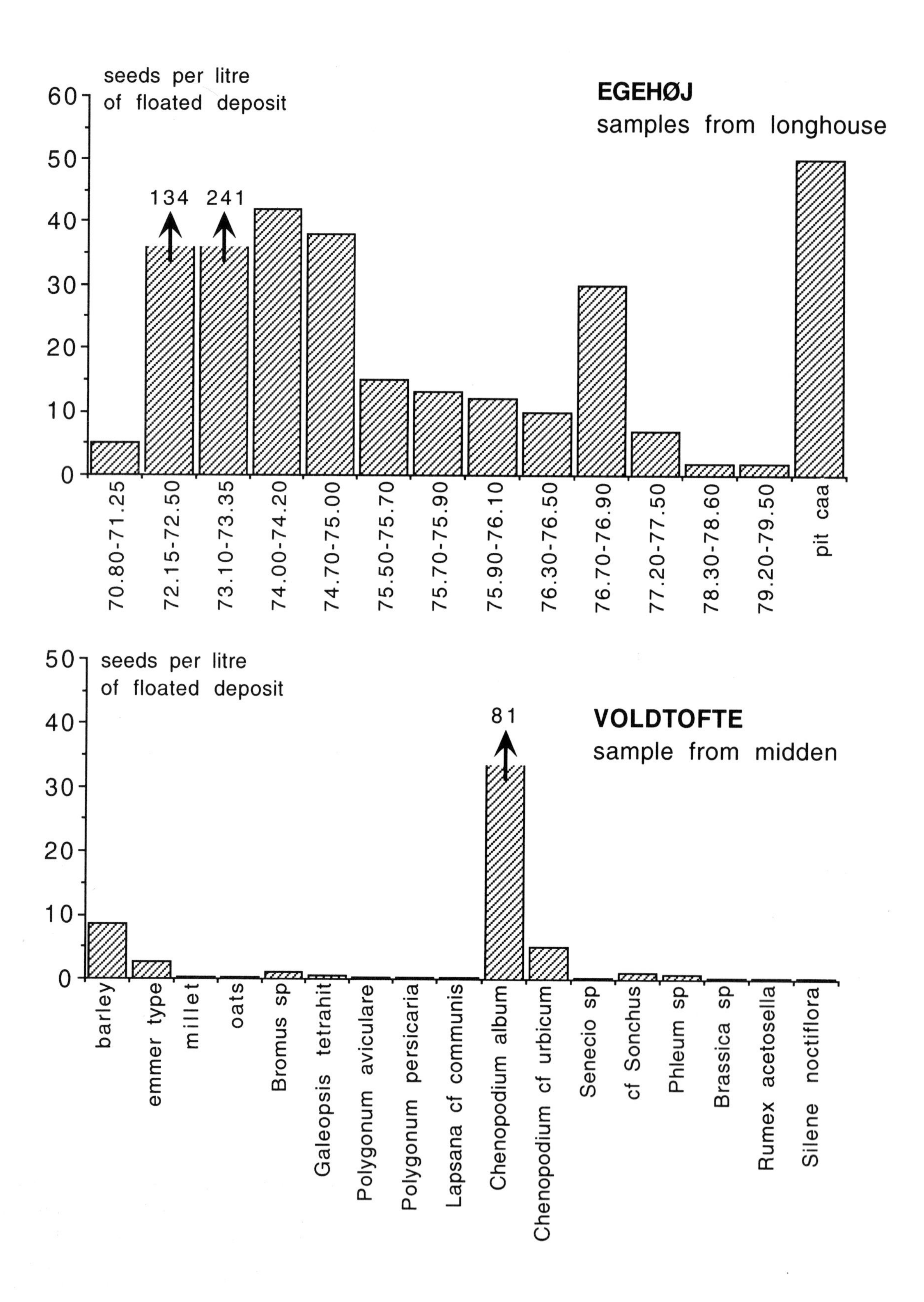

Figure 5.2: Seed frequency at the sites of Egehøj and Voldtofte. Note that the horizontal axes are dissimilar: for Voldtofte it gives the plant taxa, while for Egehøj it gives the co-ordinates of the various samples from within the longhouse.

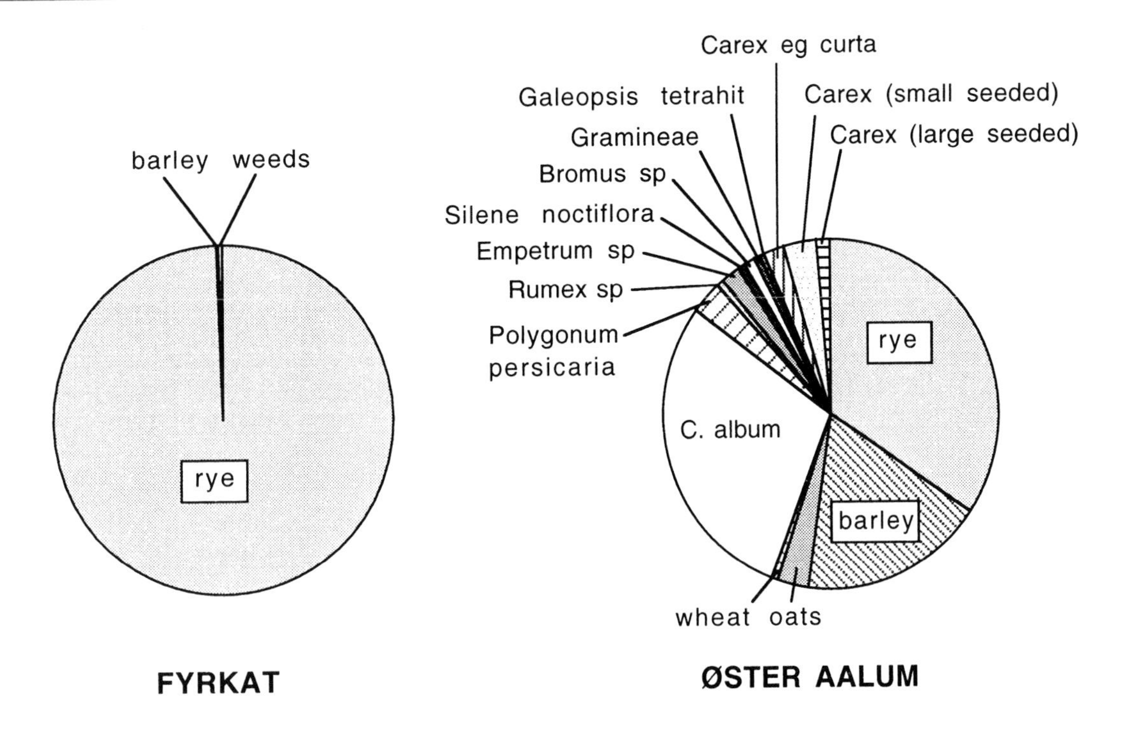

Figure 5.3: Contents of botanical samples from two Danish Viking age sites. Fyrkat fromHelbæk (1970, 1974), Øster Aalum from Rowley-Conwy (1988, table 1).

Storage in the roof space of course implies that this area was at least partly floored in. It is difficult to prove categorically that such floors existed, although concentrations of posts in some longhouses have been argued to indicate raised floors in part of the structures (Tringham 1971, 119). Longitudinal rows of interior posts were certainly necessary for technological reasons (Startin 1978). It would therefore be relatively straightforward to add an upper floor; the structure was certainly robust enough to bear the extra weight. Startin (op. cit., 150) even suggests that there could have been two floors in the roof space of LBK longhouses. Certainly, cereal storage on an upper floor is commonly assumed (ibid., 145; Tringham, op. cit.).

The foregoing applies only to rectangular houses. It is unclear how easy it would be to put an upper floor into a round house such as those built in prehistoric Britain. The structures are less robust and massive: roundhouses present no flat surface to the wind (Reynolds 1979, 35), while longhouses do and were apparently reinforced to cope with this (Startin 1978, 149). The absence of rows of interior posts would make the insertion of storage space in the roof more difficult. Reynolds (1979, 35) mentions that the central post present in some roundhouses could among other possibilities have supported an upper floor, but no reconstructions either at Butser (Reynolds op. cit.) or at Flag Fen (Pryor 1991) have attempted to include such a structure. Nothing in the house plans examined in order to build the Flag Fen house nor in any of the timbers found during the Flag Fen excavation have there been any indications of an upper floor; it would be necessary to raise the wall height to accomodate such a floor, and at least in the smaller houses this would be difficult (Maisie Taylor pers. comm.).

It therefore appears unlikely that upper floors would have been a regular feature of roundhouses in prehistoric Britain. As a result, when such houses burnt down accidentally, the huge dense grain deposits encountered in Danish and other continental longhouses were not created. This, it is suggested, is the major reason why such cereal finds are such a rare feature of the British archaeological record. This does not preclude the storage of some cereals inside roundhouses on shelving or in suspended containers. This may account for the finds of cereals inside the Iron Age roundhouse below the Roman fort at South Shields (Marijke van der Veen pers. comm.); in the Iron Age example below Peel Castle on the Isle of Man (Tomlinson in press); and in the Bronze Age example from Lairg (Carter and Holden this volume). However, the later prehistoric period in Britain is well equipped with other methods of storing cereals. The enigmatic four-posters on many settlements have been convincingly argued to be cereal storage facilities (Gent 1983), and pits would also have been suitable (Cunliffe 1984, Reynolds 1979; see discussion in Legge 1989, 220–222).

Taphonomic factors resulting from differences in build ing design may therefore be the main reason why the prehistory of Britain has few massive cereal finds. Houses would burn relatively frequently because of the fires inside them, while four-posters would not be exposed to anything like the same risk of burning. A burnt longhouse provides a data capsule recording what was in store at a particular point in time, while a four-poster hardly ever provides more than its four post holes – though some possible associations with cereals are known (Gent 1983, 249–50).

(d) Other Cultural Traditions: Germany and Japan

A brief look is here taken at two other archaeological regions in an attempt to see whether the same patterns recur. In Germany some evidence initially appears to contradict the argument put forward here, while the Japanese evidence offers support.

The major excavations in the Flögeln-Eekhölten region of Lower Saxony have produced a very large number of rectangular houses dating from the 1st to the 6th centuries AD, and many of these were destroyed by fire (Zimmerman 1992). However, only the site of Archsum-Mellenknop has a house containing much grain, a very large find (Kossack, Harck and Reichstein 1974, 325). Zimmerman (op. cit.) discusses cereal storage at some length, concluding that the roof space would be an ideal location. However, there is some documentary evidence from the *Lex Alemanni* suggesting that the roof spaces were nevertheless not used for storage, and the settlements have many five-post structures interpreted as cereal storage facilities (Zimmerman op. cit.; for a more general discussion of storage structures see Zimmerman 1991). Thus the near-absence of large cereal finds in these rectangular houses does not contradict the argument put forward above: for some reason, despite building suitable houses, the inhabitants nevertheless chose to store their cereals elsewhere.

The Japanese evidence is more parallel to that of Britain. The Yayoi period (500 BC–AD 300) sees the first major cultivation of rice in Japan. The central importance of rice to the economy is beyond doubt: several finds of paddy fields are known, and many technological items are interpreted as rice harvesting and processing equipment. Burnt houses are relatively common; however, there are few or no large samples of rice from such burnt structures. The houses in question are relatively small and more or less round. Numerous six-post structures are interpreted as rice stores (Museum of Yayoi Culture 1991). The number of Yayoi houses now excavated in Japan is very large, so the absence of major rice finds inside them forms a good parallel to the British evidence.

3. CEREAL CULTIVATION IN EARLY PREHISTORIC BRITAIN

(a) A Review of the Debate

The importance of cereals in the economy of neolithic and early bronze age Britain has been the subject of considerable debate. On the one hand, it has been argued that the importance of cereal cultivation has been overemphasised, and that wild plant foods would have been more important in the diet (Entwhistle and Grant 1989, Moffett, Robinson and Straker 1989, Greig 1991, Moffett 1991). Some strands of this view can be traced back to a brief discussion of the plant remains on three English Grooved Ware sites (M. Jones 1980); all comprised pit scatters, in which the "strong representation of woodland food plants" was stressed (op. cit., 62). Against this, it has been argued that cereals were central to the economy but that taphonomic factors usually diminish their visibility (Legge 1989, G. Jones in press).

Some arguments have been put forward minimising the importance of cereals on grounds other than the botanical remains themselves, and these will be discussed first. Entwhistle and Grant (1989, 204) point to the absence of four-posters in the earlier period. However, Gent (1983) links such structures to the centralised storage and redistribution of cereals, because more are found on fortified settlements. Organised redistribution is likely to have been less prevalent in the earlier period, and this may account for the appearance of four-posters only later in prehistory. Other forms of above-ground storage can be envisaged for the earlier periods which might leave no archaeological trace. The same authorities (Entwhistle and Grant 1989, 205) also quote the apparent "ritual" context of some neolithic ard marks (Rowley-Conwy 1987). Whether or not the marks preserved under burial mounds had a ritual function, they nevertheless testify to the *existence* of the ard in neolithic Britain. It is unlikely that ards would be used only for ritual purposes, so the ard marks can surely be taken to indicate that field preparation was normally carried out using this implement. The ritual context of some preserved examples does not support the "transient hoe-based horticulture" suggested by Grant and Entwhistle (op. cit., 208).

Arguments about the role of cereals are however mostly based on the botanical remains themselves. Continental comparisons are often made. "The direct evidence for cereal cultivation in southern England during the neolithic and earlier bronze age, is not an adequate foundation for its ready acceptance as a central feature in the economy" (Entwhistle and Grant 1989, 204). "A comparison with the cereal evidence from *Linearbandkeramik* sites in northern Europe further highlights the paucity of the British evidence... *Linearbandkeramik* sites are more comparable to some of the English iron age sites that those of the neolithic and early bronze age" (ibid.). The response by Legge (1989) makes three main points: (1) large scale flotation has been rare; (2) most of the earlier sites are of a ritual nature; (3) the samples that are available appear to be the waste products from crop processing. The thoughtful review of the evidence by Moffett, Robinson and Straker (1989) puts some of this debate into perspective. Twenty four neolithic and 2 beaker sites are listed where flotation (usually small-scale) has been used. Not all of these sites are obviously ceremonial in nature. Despite this, they list only 922 cereal grains from the 26 sites, an average of 35.5 cereal grains per site, so the conclusion that "neolithic charred plant remains have the reputation of being sparse" (op. cit., 243–45) seems a conservative statement! They agree that many of the remains represent crop processing residues (op. cit., 245), but conclude that wild plant resources are likely to have played a much more important role in neolithic Britain than in the LBK (op. cit., 252–255).

(b) Sites, Seeds and Diet

It will be argued here that taphonomic factors are probably responsible for the apparent under-representation of cereals in the neolithic and early bronze age. The first point to make is that the visible differences between Britain and Denmark discussed in the preceding section are much greater than the differences between the early and late periods in prehistoric Britain; if taphonomic factors can be successfully invoked to explain the former, they may be expected to cast light on the latter as well. It is also worth remembering that our views of iron age agriculture in the north of England have changed relatively recently. Piggott (1958) envisaged "the largely uncultivated lands of Brigantia" (op. cit., 16) populated by "Celtic cow-boys and shepherds, footloose and unpredictable, moving with their animals over rough pasture and moorland" (ibid., 25). This view was reiterated as recently as 1978 by Frere, who believed that the Roman army "undoubtedly had the effect of introducing arable cultivation" (Frere 1978, 310). This picture has been comprehensively disproved by a number of recent pieces of work: (a) systematic botanical work has recovered numerous finds of cereals (van der Veen 1992, Huntley and Stallibrass 1995); (b) extensive pre-Roman field systems have been identified, comprising both cord rig and terrace systems (Topping 1989, Mercer and Tipping 1994); (c) the remains of domestic animals are numerous (Huntley and Stallibrass op. cit.); and (d) a major iron age erosional event has been linked to agricultural intensification (Mercer and Tipping 1994). There is therefore no doubt whatever that iron age subsistence in the north definitely included a very major cereal component; as Topping (1989, 175) aptly remarks, "we now have to re-evaluate our theories and accept that the footloose Celtic cowboy has finally met his High Noon." The fact that this remained unknown until so recently should warn us against dismissing neolithic cultivation on the basis of negative evidence.

Site function is clearly crucial to the debate. Determination of whether a feature was ceremonial or domestic in nature is not always easy, however. The three Grooved Ware pit scatters mentioned above that provided the samples discussed by M. Jones (1980) were described as "containing what appeared to be *domestic refuse*" (op. cit., 61, added emphasis). One of these pit scatters is however that from Firtree Field, Down Farm. The full publication of this site makes interesting reading: the pits are identified as *non-domestic* on virtually the full suite of their contents, including animal bones (Legge 1991, 68, 72), stone tools (Brown 1991, 111–114) and pottery (Cleal 1991, 140–141). This unusual agreement among specialists does not, however, extend as far as the plant remains: in the final publication the charred items are listed pit by pit without discussion of possible "ritual" connotations (M. Jones 1991), and it may be that structured deposition is less visible with regard to plant remains that to the other classes of material from the pits. The other English Grooved Ware sites considered by M. Jones (1980), Mount Farm and Barton Court Farm, were apparently similar to those at Down Farm and located close to major neolithic ceremonial sites (op. cit., 61). Could these also have had a non-domestic function? This is a question that could indeed be directed at other neolithic sites of indeterminate nature.

It is difficult to name a single definite neolithic or early bronze age settlement in England or Wales excavated prior to the review by Moffett, Robinson and Straker (1989) that has yielded large macrobotanical samples. In Scotland, however, this is not the case. To mention one example, Scord of Brouster on Shetland is a neolithic settlement comprising several houses, and has produced a substantial number of samples (Milles 1986). Two groups are plotted in Figure 5.4. Samples 79 and 82 were recovered from the west end of the floor of house 3, while samples 56, 57 and 58 came from the base of the central hearth belonging to phase 2a of house 1. The floor samples were dense and dominated by large grains of barley, while the hearth samples were more diffuse, were dominated by weed and chaff fragments, and had a larger proportion of small barley grains (all details from Milles op. cit., tables 24, 27 and 28). The floor samples are interpreted as cleaned grain, the hearth samples as processing waste (op. cit., 121). Thus we have an unambiguous neolithic settlement, producing many samples of both cleaned grain and waste — in association with a 2.5 ha field system. This demonstrates that cereals provided the overwhelming majority of the plant foods eaten by the neolithic inhabitants of Scord of Brouster – and this in an area which can scarcely be described as the best agricultural land in Britain.

How likely are nuts, fruits and tubers to have played a more important role than cereals in neolithic subsistence further south? Tubers have been mentioned by Moffett (1991) and Murphy (1989). Most such resources are not energetically efficient, however, as their calorific yield is relatively low and the labour involved in digging them up is considerable (for a discussion in a mesolithic context see Rowley-Conwy 1986, 27). Fruits have very low calorific values, and many are difficult to store. The role of hazel nuts thus becomes crucial, as the high calorific value and the storability of the nuts mark them out as about the only wild plant likely to compete with the cereals. Legge stresses that hazel nut shell is a waste product with

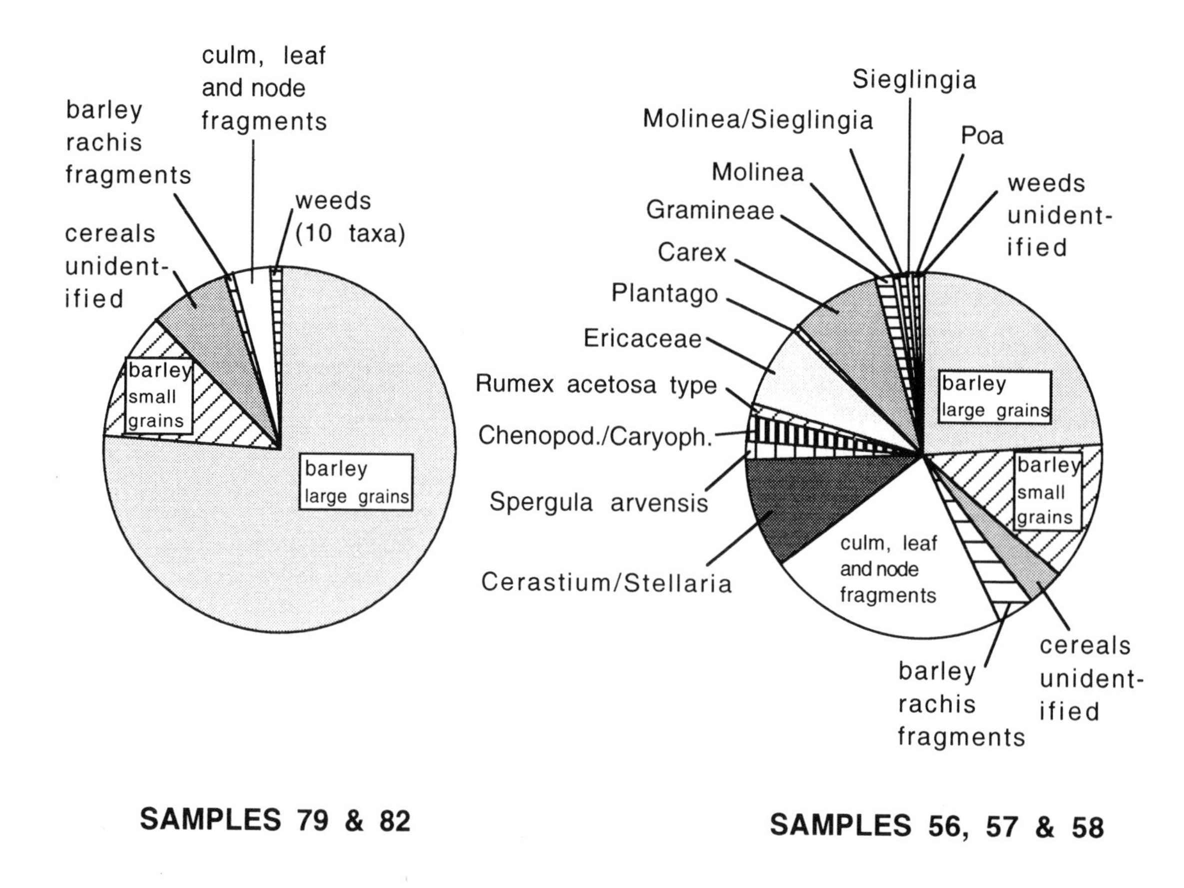

Figure 5.4: Amalgamated contents of botanical samples from Scord of Brouster. Samples 79 and 82 from Milles (1986, fiche table 28), samples 56,57 and 58 from ibid. (fiche table 27). The division between large and small barley grains is placed at 4mm in length (ibid., fiche p. 220).

high survival potential when charred, and is therefore likely to be over-represented in the archaeological record (op. cit., 218–19). The present writer is therefore unconvinced that wild plant foods *ever* formed more than a minor part of the diet. To be sure, they may well have played a more important role in the neolithic than they did in the iron age, but this does not mean that they were ever more than a supplement. The viability of a "half and half" economy in temperate Europe is in some doubt, given the scheduling problems that would occur in autumn; a relatively rapid switch to a predominantly agricultural economy is more likely (Zvelebil and Rowley-Conwy 1986, 80).

(c) The Continental Comparison

Continental comparisons have been brought into the debate, usually in the form of the central European LBK culture (see above). There is disagreement as to just how major the LBK evidence actually is (Legge 1989, 221 *contra* Moffett, Robinson and Straker 1989, 254–255 and Entwhistle and Grant 1989, 204). A review of the LBK evidence would be helpful, particularly in view of the rectangular houses used by this culture. If the more spectacular finds come from burnt longhouses, such finds would (as in Denmark – see above) be highly visible and easy to recover.

Other continental areas are also relevant, however. Denmark has evidence of cereal cultivation from the start of the early neolithic in the form of pottery impressions (Helbæk 1954b); however, charred evidence is scanty in the extreme through most of the neolithic. The oldest major find, 7 litres of cereals, comes from Birknæs (Helbæk 1952), a find now dated to the late neolithic (Jørgensen 1979). Acorns and apples have been recovered from the later part of the middle neolithic (Jørgensen 1977); Denmark could thus be argued to resemble Britain in its scarcity of direct evidence from the early and middle neolithic. The crucial exception is the causewayed camp at Sarup, dating from the beginning of the middle neolithic. Large scale flotation was carried out by the writer, and numerous finds of virtually pure emmer wheat are known (Jørgensen 1976, 1981; Rowley-Conwy in prep.). This site emphasises the importance of cereal cultivation at a time when it is otherwise rather scantily attested; Sarup thus plays a similar role to the site of Hambledon Hill in Britain (Legge 1989, 218–19). The Danish situation is due to a near-total absence of flotation on neolithic sites.

Southern Sweden presents an even more extreme picture. Throughout the neolithic, pottery impressions form the bulk of the evidence, with a find of charred cereals, weeds and fruits coming from the site of Piledal (Hjelmqvist 1985, L. Larsson 1992, M. Larsson 1984). A quite remarkable piece of evidence testifying to significant cereal cultivation is however available from the pile-dwelling at Alvastra, dating from the start of the middle neolithic. Pollen core I from Alvastra went straight though the culture layer, and at this point pollen values of Cerealia suddenly increase to *over 80% of the non-arboreal pollen*. "It is supposed that the cereal pollen was released when the ears were threshed on the floor of the pile dwelling" (Göransson 1988, 65). Pollen cores II and III, only 15 m north and 32 m northeast respectively, showed only minor peaks below 10% of NAP (Göransson 1988), a clear indication that cereal pollen does not travel very far at all. Thus the importance of cereal cultivation in the southern Swedish neolithic cannot be doubted – but the charred evidence for this is no more convincing than that from Britain.

(d) Evidence for Major Cereal Cultivation in the British Neolithic

The debate has now got to the point where new evidence rather than new arguments is the way forward. The best resolution to the debate would be the finding of rectangular neolithic houses in Britain, burnt down and preserving major finds of cereals. Two such sites are now known, both of them appearing since the review of the neolithic evidence published by Moffett, Robinson and Straker (1989).

Balbridie, near Banchory in Scotland, comprises a rectangular building measuring 24 x 12 metres. A total of 14 radiocarbon dates, some on the plant remains themselves, indicate a date in the early 3rd millennium BC (uncalibrated). The building was destroyed by fire; in the destruction layers were very large numbers of cereal grains, of which a sample of some 20,000 has been identified. Emmer, barley and bread wheat grains predominate; there are few spikelet fragments and relatively few weeds. A small deposit of flax seeds was found. Among collected foodstuffs, apple and hazel nut are mentioned. Above-ground storage within the building is the likely origin of the material (all information from Fairweather and Ralston 1993).

Lismore Fields, near Buxton, comprises two rectangular buildings which burnt down. A radiocarbon date on some of the seeds indicates a date at the very start of the 3rd millennium BC (uncalibrated). The large sample of cereals from the destruction layer consists almost entirely of emmer wheat. Along with the grains, considerable quantities of chaff were also recovered. A deposit of flax seeds was also found. Among gathered plants, hazel nut shells were common, and in addition apples were found. It is suggested that the nutshell may have been stored as kindling (all information from G. Jones in press).

These two major neolithic finds provide evidence more spectacular than that from later periods in Britain, and go some way towards putting the British neolithic on a par with the major continental finds mentioned above. Both contain hazel nuts and apples, but they offer no support for the view that such wild foods were more important than the cereals. They are strong evidence that cereal cultivation was indeed the predominant source of plant foods in the British neolithic, and as such they also strongly suggest that the low visibility of cereal cultivation elsewhere is indeed the result of taphonomic factors.

4. CONCLUSION

The conclusion of this paper is that taphonomic factors affect evidence of plant cultivation a great deal more than is often realised, and at many levels. The taphonomic pathways of cereals between being harvested in prehistory and being published by archaeobotanists are highly varied; the likelihood of a species being preserved by charring is dependent on a whole host of factors, many of which introduce distortions and biases. In addition, it is argued here that storage method and building morphology will directly affect the chances of major cereal stores being burnt and thus preserved. This is turn will strongly affect visibility of these plants in the archaeological record. Finally, archaeological visibility will be exacerbated if flotation is used to differing degrees in different areas. The standard British practice of floating very small soil samples is virtually guaranteed to recover very small botanical samples. Until large-scale flotation is used as a matter of course, neolithic agriculture in Britain will continue to be little known.

ACKNOWLEDGEMENTS

I would like to thank Helena Hamerow for providing me with the references to the Flögeln-Eekhölten project, and Akira Matsui and Simon Kaner for translating the Japanese reference for me. I am grateful to Maisie Taylor (Fenland Archaeological Trust) for information about roundhouse construction. Thanks to Glynis Jones, Philippa Tomlinson and Marijke van der Veen for permission to use their unpublished results, and to Tony Legge for reading an earlier version of this paper and offering me his comments. None of the foregoing should be held responsible for any of the contents of the paper.

BIBLIOGRAPHY

Berglund, J. 1982. Kirkebjerget – a late bronze age settlement at Voldtofte, south-west Funen. *Journal of Danish Archaeology* 1, 51–63.

Boas, N.A. 1983. Egehøj. A settlement from the early bronze age in east Jutland. *Journal of Danish Archaeology* 2, 90–101.

Brown, A. 1991. Structured deposition and technological change among the flaked stone artefacts from Cranbourne Chase. In: *Papers on the Prehistoric Archaeology of Cranbourne Chase*, eds. J. Barrett, R. Bradley and M. Hall, 101–133. Oxbow Monograph 11. Oxford: Oxbow Books.

Clark, J.G.D. 1952. *Prehistoric Europe. The Economic Basis*. Cambridge: University Press.

Cleal, R.M.J. 1991. Cranbourne Chase – the earlier prehistoric pottery. In: *Papers on the Prehistoric Archaeology of Cranbourne Chase*, eds. J. Barrett, R. Bradley and M. Hall, 134–200. Oxbow Monograph 11. Oxford: Oxbow Books.

Cunliffe, B. 1984. *Danebury: an Iron Age Hillfort in Hampshire* (2 vols). London: Council for British Archaeology, research report 52.

Entwistle R. and Grant A. 1989. The evidence for cereal cultivation and animal husbandry in the southern British neolithic and bronze age. In: *The Beginnings of Agriculture*, eds. A. Milles, D. Williams and N. Gardner, 203–215. British Archaeological Reports, International Series 496. Oxford: BAR.

Fairweather, A.D. and Ralston, I.B.M. 1993. The neolithic timber hall at Balbridie, Grampian Region, Scotland: the building, the date, the plant macrofossils. *Antiquity 67*, 313–323.

Frere, S. 1978. *Britannia. A History of Roman Britain* (revised edition). London: Routledge and Kegan Paul.

Gent, H. 1983. Centralized storage in later prehistoric Britain. *Proceedings of the Prehistoric Society* 49, 243–267.

Greig, J.R.A. 1991. The British Isles. In *Progress in Old World Palaeoethnobotany* eds. W. van Zeist, K. Wasylikowa and K.-E. Behre, 299–334. Rotterdam: Balkema.

Hatt, G. 1931. En brandtomt af et jernalderhus paa Mors (French summary). *Aarbøger for Nordisk Oldkyndighed og Historie* 1931, 83–119.

Hatt, G. 1937. *Landbrug i Danmarks Oldtid*. Copenhagen: Udvalget for Folkeoplysnings Fremme.

Hatt, G. 1941. Forhistoriske plovfurer i Jylland (French summary). *Aarbøger for Nordisk Oldkyndighed og Historie* 1941, 155–165, XVI-XVII.

Helbæk, H. 1938. Planteavl. pp 216–226 of: Jernalders bopladser i Himmerland (by G. Hatt), *Aarbøger for Nordisk Oldkyndighed og Historie* 1938, 119–266.

Helbæk, H. 1952a. Early crops in southern England. *Proceedings of the Prehistoric Society* 18, 194–233.

Helbæk, H. 1952b. Spelt (*Triticum spelta* L.) in bronze age Denmark. Acta Archaeologica 23, 97–107.

Helbæk, H. 1954a. Prehistoric food plants and weeds in Denmark. A survey of archaeobotanical research 1923–1954. *Danmarks Geologisk Undersøgelse* series II, 80, 250–261.

Helbæk, H. 1954b. Store Valby – kornavl i Danmarks første neolitiske fase (English summary). *Aarbøger for Nordisk Oldkyndighed og Historie* 1954, 198–204.

Helbæk, H. 1970. Da rugen kom til Danmark (English summary). *Kuml* 1970, 279–296.

Helbæk, H. 1974. The Fyrkat grain. A geographical and chronological study of rye. In *Fyrkat*, by O. Olsen, E. Roesdahl and H.W. Schmidt, 1–41. Copenhagen: Nordiske Fortidsminder series B, 2.

Hillman, G. 1981. Reconstructing crop husbandry practices from charred remains of crops. In *Farming Practice in British Prehistory*, ed. R. Mercer, 123–162. Edinburgh: University Press.

Hillman, G. 1984. Interpretation of archaeological plant remains: the application of ethnographic models from Turkey. In *Plants and Ancient Man*, eds. W. van Zeist and W.A. Casparie, 4–41. Rotterdam: A.A. Balkema.

Hjelmqvist, H. 1985. Economic plants from two stone age settlements in southernmost Scania. *Acta Archaeologica* 54 (for 1983), 57–63.

Huntley, J.P. and Stallibrass, S. 1995. *Plant and Vertebrate Remains from Archaeological Sites in Northern England: Data Reviews and Future Directions*. Durham: Architectural and Archaeological Society of Durham and Northumberland. (Research Report 4).

Jæger, A. and Laursen, J. 1983. Lindebjerg and Røjle Mose. Two early bronze age settlements on Fyn. *Journal of Danish Archaeology* 2, 102–117.

Jarman, H.N., Legge, A.J. and Charles, J.A. 1972. Retrieval of plant remains from archaeological sites by froth flotation. In *Papers in Economic Prehistory*, ed. E.S. Higgs, 39–48. Cambridge: University Press.

Jessen, K. 1933. Planterester fra den ældre jernalder i Thy (German summary). *Botanisk Tidsskrift* 42, 257–288.

Jones, G. 1984. Interpretation of plant remains: ethnographic models from Greece. In *Plants and Ancient Man*, eds. W. van Zeist and W.A. Casparie, 43–61. Rotterdam: A.A. Balkema.

Jones, G. in press. Charred plant remains from the neolithic settlement at Lismore Fields, Buxton. To appear in a volume on Lismore Fields, ed. D. Garton, to be published by the Prehistoric Society.

Jones, M. 1980. Carbonised cereals from Grooved Ware contexts. *Proceedings of the Prehistoric Society* 46, 61–63.

Jones, M. 1991. Down Farm, Woodcutts: the carbonised plant remains. In: *Papers on the Prehistoric Archaeology of Cranbourne Chase*, eds. J. Barrett, R. Bradley and M. Hall, 49–53. Oxbow Monograph 11. Oxford: Oxbow Books.

Jørgensen, G. 1976. Et kornfund fra Sarup (English summary). *Kuml* 1976, 47–64.

Jørgensen, G. 1977. Acorns as a food-source in the later stone age. *Acta Archaeologica* 48, 233–238.

Jørgensen, G. 1979. A new contribution concerning the cultivation of spelt, *Triticum spelta L.*, in prehistoric Denmark. *Archaeo-Physika* 8, 135–145.

Jørgensen, G. 1981. Korn fra Sarup (English summary). *Kuml* 1981, 221–231.

Klindt-Jensen, O. 1951. Freds- og krigstid i Bornholms jernalder. *Fra Nationalmuseets Arbejdsmark* 1951, 16–23.

Kossack, G., Harck, O. and Reichstein, J. 1974. Zehn Jahre Siedlungsforschung in Archsum auf Sylt. *Bericht der Römisch-Germanischen Kommission* 55, 261–427.

Larsson, L. 1992. Settlement and environment during the middle neolithic and late neolithic. In *The Archaeology of the Cultural Landscape* eds. L. Larsson, J. Callmer and B. Stjernqvist, 91–159.

Larsson, M. 1984. *Tidigneolitikum i Sydvästskåne. Kronologi och Bosättningsmönster*. Acta Archaeologica Lundensia, series in 4°, no. 17. Lund: Gleerup.

Legge, A.J. 1989. Milking the evidence: a reply to Entwistle and Grant. In: *The Beginnings of Agriculture*, eds. A. Milles, D. Williams and N. Gardner, 217–242. British Archaeological Reports, International Series 496. Oxford: BAR.

Legge, A.J. 1991. The animal remains from six sites at Down Farm, Woodcutts. In: *Papers on the Prehistoric Archaeology of Cranbourne Chase*, eds. J. Barrett, R. Bradley and M. Hall, 54–100. Oxbow Monograph 11. Oxford: Oxbow Books.

Mercer, R. and Tipping, R. 1994. The prehistory of soil erosion in the northern and eastern Cheviot Hills, Anglo-Scottish borders. In: *The History of Soils and Field Systems*, ed. S. Foster and T.C. Smout, 1–25. Aberdeen: Scottish Cultural Press.

Milles, A. 1986. Charred remains of barley and other plants from Scord of Brouster. In: *Scord of Brouster. An Early Agricultural Settlement on Shetland*, by A. Whittle et al., 119–122. Oxford: Oxford University Committee for Archaeology, Monograph 9.

Moffett, L. 1991. Pignut tubers from a bronze age cremation at

Barrow Hills, Oxfordshire, and the importance of vegetable tubers in the prehistoric period. *Journal of Archaeological Science* 18, 187–191.

Moffett L., Robinson M.A. and Straker V. 1989. Cereals, fruits and nuts: charred plant remains from neolithic sites in England and Wales and the neolithic economy. In: *The Beginnings of Agriculture*, eds. A. Milles, D. Williams and N. Gardner, 243–261. British Archaeological Reports, International Series 496. Oxford: BAR.

Murphy, P. 1989. Carbonised neolithic plant remains from The Stumble, an intertidal site in the Blackwater Estuary, Essex. *Circaea* 6(1), 21–38.

Museum of Yayoi Culture, 1991. *Yoyoi Culture: In Pursuit of the Origins of Japanese Culture* (in Japanese). Tokyo: Heibonsha.

Piggott, S. 1958. Native economies and the Roman occupation of North Britain. In: *Roman and Native in North Britain*, ed. I.A. Richmond, 1–27. London: Nelson.

Pryor, F. 1991. *Flag Fen*. London: Batsford.

Reynolds, P.J. 1979. *Iron Age Farm. The Butser Experiment*. London: British Museum.

Rowley-Conwy, P. 1978. Forkullet korn fra Lindebjerg. En boplads fra ældre bronzealder (with full English translation: The carbonised grain from Lindebjerg). *Kuml* 1978, 159–171.

Rowley-Conwy, P. 1982. Bronzealder korn fra Voldtofte (with full English translation: A new sample of carbonized grain from Voldtofte). *Kuml* 1982–3, 139–152.

Rowley-Conwy, P. 1984. The Egehøj cereals. Bread wheat (*Triticum aestivum* s.l.) in the Danish early bronze age. *Journal of Danish Archaeology* 3, 1984, 104–110.

Rowley-Conwy, P. 1986. Between cave painters and crop planters: aspects of the European mesolithic. In: *Hunters in Transition*, ed. M. Zvelebil, 17–32. Cambridge: University Press.

Rowley-Conwy, P. 1987. The interpretation of ard marks. *Antiquity* 61, 263–266.

Rowley-Conwy, P. 1988. Rye in Viking age Denmark: new information from Øster Aalum, North Jutland. *Journal of Danish Archaeology* 7, 182–190.

Startin, W. 1978. Linear Pottery Culture houses: reconstruction and manpower. *Proceedings of the Prehistoric Society* 44, 143–159.

Tomlinson, P.R. (in press). The charred cereal deposit. In: *Excavations at Peel Castle 1982–87* ed. D.S. Freke et al. Liverpool: University Press.

Topping, P. 1989. Early cultivation in Northumberland and The Borders. *Proceedings of the Prehistoric Society* 55, 161–179.

Tringham, R. 1971. *Hunters, Fishers and Farmers of Eastern Europe*. London: Hutchinson.

van der Veen, M. 1992. *Crop Husbandry Regimes*. Sheffield: J.R. Collis Publications, Department of Archaeology and Prehistory (Sheffield Archaeological Monographs 3).

Zimmerman, W.H. 1991. Erntebergung in Rutenberg und Diemen aus archäologischer und volkskundlicher Sicht. *Néprajzi Értesitö a Néprajzi Muzeum Évkönyve* 71–73 (for 1989–91), 71–104.

Zimmerman, W.H. 1992. *Die Siedlungen des 1. bis 6. Jahrhunderts nach Christus von Flögeln-Eekhölten, Niedersachsen: Die Bauformen und ihre Funktionen*. Probleme der Küstenforschung im Südlichen Nordseegebiet 19. Hildesheim: August Lax 1992.

Zvelebil, M. and Rowley-Conwy, P. 1986. Foragers and farmers in Atlantic Europe. In: *Hunters in Transition*, ed. M. Zvelebil, 67–93. Cambridge: University Press.

6. Otter (*Lutra lutra L.*) spraint: an investigation into possible sources of small fish bones at coastal archaeological sites

Rebecca A. Nicholson

SUMMARY

Accumulations of small fish bones have been examined from three archaeological sites of prehistoric date on the Orkney Isles, and from two groups of contemporary otter spraint from Northern Scotland and Shetland. By examining the species present, the proportional representation of skeletal elements, and the amount and types of damage to bones, it is concluded that each of the three archaeological groups of small bones were deposited by otters, in their spraint, and not by the actions of humans or other predators, nor by environmental mechanisms.

1. INTRODUCTION

Background

Increasingly, as sampling and sieving become standard procedures for archaeological excavations, small fish bones are recovered. Determining the origin of these bones – whether deposited as human refuse, as a "ritual deposit", or by "natural" mechanisms such as in animal excreta, or accumulated by wind or water is not easy, especially for exposed coastal sites where many mechanisms are possible.

Alwyne Wheeler, when analysing a large deposit of small bones from the Neolithic chambered cairn at Quanterness, Orkney, considered the possibility of otters having been responsible, but concluded that a "ritual" explanation is more probable (Wheeler 1979, 148). His argument rested on the unlikelihood of otters sprainting within their den or holt, an argument refuted by Colley (1987, 132–3) on the advice of zoologists with a particular knowledge of otters. Colley (1987) concluded that concentrations of small fish bones at the Iron Age site of Bu Broch, Orkney, and the Neolithic chambered cairn at Isbister, Orkney (Colley 1983) originated as otter spraint. She noted the similarity of the species represented in the archaeological material with those regularly captured by coastal otters. I was given the opportunity to examine large concentrations of mainly small (from fish of about 100–200 mm.) and tiny (fish of less than 100 mm.) fish bone from three Orkney sites: from excavations of Iron Age structures and deposits at Pool, and of an Iron Age round house and associated features at Tofts Ness, both on Sanday (Hunter et al, forthcoming) and from excavations of a Neolithic compartmented structure at the Links of Noltland, Westray (Clarke unpublished). In each case the accumulations of bone could have had either a "cultural" origin (fish placed there by humans) or a "natural" origin (deposited by agencies other than mankind). Of the "natural" mechanisms, otters were likely considering the large size of the bone deposits and their position in association with archaeological monuments. At Tofts Ness the bone accumulation was related stratigraphically to the period of round house abandonment (S.Dockrill, pers. comm.). That otters were around is demonstrated by the discovery of otter bones at Pool and Tofts Ness in the Iron Age phases (J.Bond, pers. comm.). The challenge was therefore to develop a methodology by which otter spraint could be recognised unequivocally in archaeological material, so that in the future similar deposits can be rapidly evaluated and, if shown to be spraint, could be largely ignored in the subsequent analysis of the human exploitation of fish.

Otter Spraints

Unusually for Britain today, otters may be still be seen along the coasts of Northern Scotland and the Northern Isles. It is likely that they were much more numerous in the past. Otter spraints typically comprise the undigested remains of prey coated with a tarry mucus. For coastal otters, fish remains account for almost the entire spraint, as fish make up 70–95% of prey (Chanin 1987, 17).

Spraints are typically 50–60 mm. long and about 10mm. in diameter, although some may be smaller and others amorphous. Otters mark notable spots with spraints; these sprainting sites may include prominent rocks and boulders, the junctions of paths, and, particularly dens and rolling places (Mayson and MacDonald 1986, 31; Chanin 1987). Spraint may also accumulate within dens or holts when young otters are being fed, although more frequently piles build up in the entrances, as the young tend to defecate outside when weaned (Chanin 1987). Fish may be left at the holt entrance and moved inside by the young.

Studies into the diet of coastal otters (Earlinge 1967, 1968; Mayson and MacDonald 1986) have shown that most of the fish taken are small and long-bodied, such as small saithe (*Pollachius virens*), rocklings (especially *Ciliata* and *Gaidropsarus* sp(p.)), the viviparous blenny (*Zoarches viviparus*) and butterfish (*Pholis gunnellus*), as well as cottids (Cottidae), small wrasse (Labridae) and lumpsuckers (*Cyclopterus lumpus*). Chanin (1987) cites figures of 57% long-bodied fish, 15% slow, bottom-dwelling fish, 8% flatfish, 7% open water fish, 12% crabs and 1% birds for the typical diet of coastal otters. Where freshwater water-bodies are available, coastal otters will catch eels (*Anguilla anguilla*) in the summer months. Rather more birds, anurans and small mammals are eaten by riverine otters. Although otters usually catch fish of around 150–300 mm. in length, smaller and larger fish are also captured and otters have been seen catching fish of an estimated 10 kg (Chanin 1987, 18). Otters will also scavenge fish corpses thrown up onto the beach.

2. MATERIALS AND METHODS

Modern Otter Spraints

For comparative purposes, a number of otter spraints from the northern coast of Scotland (Freswick, Caithness) and from Shetland were examined by the author. In total, 6 spraints from Freswick and 9 from Shetland were diss-agreggated manually in a gentle stream of water and air-dried. The fish remains were identified by skeletal element and, where possible, to species, with the aid of a dissecting microscope and the extensive fish bone reference collection held in the Environmental Archaeology Unit, University of York. Further spraints were dissaggregated and rapidly scanned to check that the chosen 15 spraints were typical in terms of the species and sizes of fish and the parts of the body represented.

Once the taxonomic identifications had been recorded, the bones were treated as one unit, irrespective of species, for analytical purposes. The proportional representation of skeletal elements was calculated as a means of determining whether similar skeletal parts were present in all the assemblages, using the method given by Dodson and Wexlar (1979, table 1), which is based on the number of bones surviving compared with the expected number:

$$PR = \frac{FO}{(FT \times MNI)} \times 100$$

where: PR = proportional representation.
FO = the number of each element recovered in an assemblage.
FT = the expected number of the element in one individual.
MNI = the minimum number of individuals, by the more frequent side of the most frequent paired bone. The figure is a minimum, as the most commonly represented bone will probably have suffered some loss too.

Because, for simplicity, the MNI has been calculated using paired bones only, it is possible that some other bones (e.g. vertebrae) may be better represented and so score over 100%. Table 6.2, which details the proportional representation, includes pooled figures for all the spraints in each of the two modern groups. The MNI has been obtained from each entire assemblage rather than by summing the individual MNI's for each spraint. This will inevitably grossly under-estimate the true MNI, but the method has been adopted to equate with archaeological assemblages, where individual spraints may not be distinguished.

Only those bones potentially identifiable to at least family level were considered, so skull fragments, hyal bones, branchials, ribs, rays, spines and small bones of the pelvic, orbital and facial regions were excluded. Elements were also only included if they could be identified from a wide range of species.

All bones were examined for signs of chewing and acid dissolution – the latter typified by pitting and polishing of the bone surfaces and edges, and "crenelation" of the vertebral articulating facets (Nicholson 1991, 1993). The proportion of these damaged bones was calculated for each assemblage. Any calcareous deposit surrounding the bones was also noted when present.

Archaeological Material

Sub-samples of 76g and 23.5g were taken from lenses of small fish bone recovered by wet-sieving to 1 mm. from context 0431/0437 from Pool and 0619 from Tofts Ness, respectively. At Pool, the fish bone accumulation was located on a rectangular flagged floor, which may have been enclosed. An ogham-marked orthostat stood towards the centre of the same floor. At Tofts Ness the lens of small fish bones came from the annexe outside the entrance to a round house. The sub-samples were obtained using a standard sample dividing box, to randomise species and skeletal elements, and comprised about one third of the deposit at Pool and one thirtieth of the much larger fish-bone lens at Tofts Ness. Small fish bones formed the majority of a considerable assemblage of fish remains from the Links of Noltland, and to date the majority of the material has only been rapidly scanned as an "assessment". Three sub-samples of 30g each from three different contexts (HO86–64, HP90–34, HQ89–31) were studied

Table 6.1: Species representation in the contemporary spraint and archaeologically-recovered assemblages of small bones.

	CONTEMPORARY		ARCHAEOLOGICAL				
Taxonomic identification	Shetland	Freswick	Pool	Tofts	Links of Noltland		
					H086-64	HP90-34	HQ89-31
Eel		+++	+	++	++	++	++
Conger eel*					+		
Salmonid indet		+				+	
Cod*		++			+		
Saithe	++		++	++			
Poor cod			+				
?Bib			+				
Five-bearded rockling	+++	+++	++	+++			
Three-bearded rockling				++			
Rockling indet		+	++		+++	++	+
Gadid indet.*	+++				+++	++	+++
Viviparous blenny	+++	+++	++	++	++		++
Butterfish	+++	++		++	+	++	
Sea-scorpion				++			
Bullhead				++			
Cottid indet	+		++	+++	+++	++	+++
3-spined stickleback	++	++	+		+		
Snake-blenny		+	+				
Ballan wrasse				+			
Corkwing wrasse				++			
Wrasse indet	+		+		++	++	++
Dab				+			
Pleuronectidae*			++			++	+
Flatfish indet.*	+						+
Anuran		+					
Small mammal		+					
Mammal							+

+ = Rare, ++ = Frequent, +++ = Abundant

* = some or most bones from fish of over 300 mm.

in more detail, but taxonomic identifications were frequently less exact than would be possible given a fuller investigation – many bones were identified to family level only (Nicholson and Jones 1992).

For all the assemblages, the proportional representation of skeletal elements was calculated as discussed above for the modern material. The numbers of major skeletal elements falling into each of the four completeness categories was recorded, and the proportions of chewed or crushed bones noted for each assemblage.

3. COMPARISONS OF MODERN OTTER SPRAINTS WITH ARCHAEOLOGICAL MATERIAL

As indicated in Table 6.1, similar fish species were represented in both of the modern spraint assemblage, and in the archaeological material. The sizes of fish were also similar; in all the assemblages bones from fish of 100–200 mm dominated. Commonly represented fish included the saithe (*Pollachius virens*), rocklings (in particular the five-bearded rockling *Ciliata mustela*), viviparous blenny (*Zoarces viviparous*), butterfish (*Pholis gunnellus*), cottids (Cottidae), 3–spined stickleback (*Gasterosteus aculeatus*), eel (*Anguilla anguilla*), wrasse (Labridae), and flatfish (mainly Pleuronectidae). Occasional species included the conger eel (*Conger conger*), snake-blenny (*Lumpenus lampretaeformis*), cod (*Gadus morhua*) and other small gadids, and tiny salmonid, probably brown trout (*Salmo trutta*). Of the cottids, sea scorpion (*Taurulus bubalis*) and bullrout (*Myoxocephalus scorpius*) were identified. Ballan wrasse (*Labrus berggylta*) and corkwing wrasse (*Crenilabrus melops*) were recognised in the material from the Links of Noltland. Frog (*Rana* sp.) or toad (*Bufo* sp.) bone was also present in some of the Freswick spraints, and the archaeological material contained fragments of mammal bone which were not observed in the modern spraint, although otters are known to occasionally prey on or scavenge mammals (Chanin 1987).

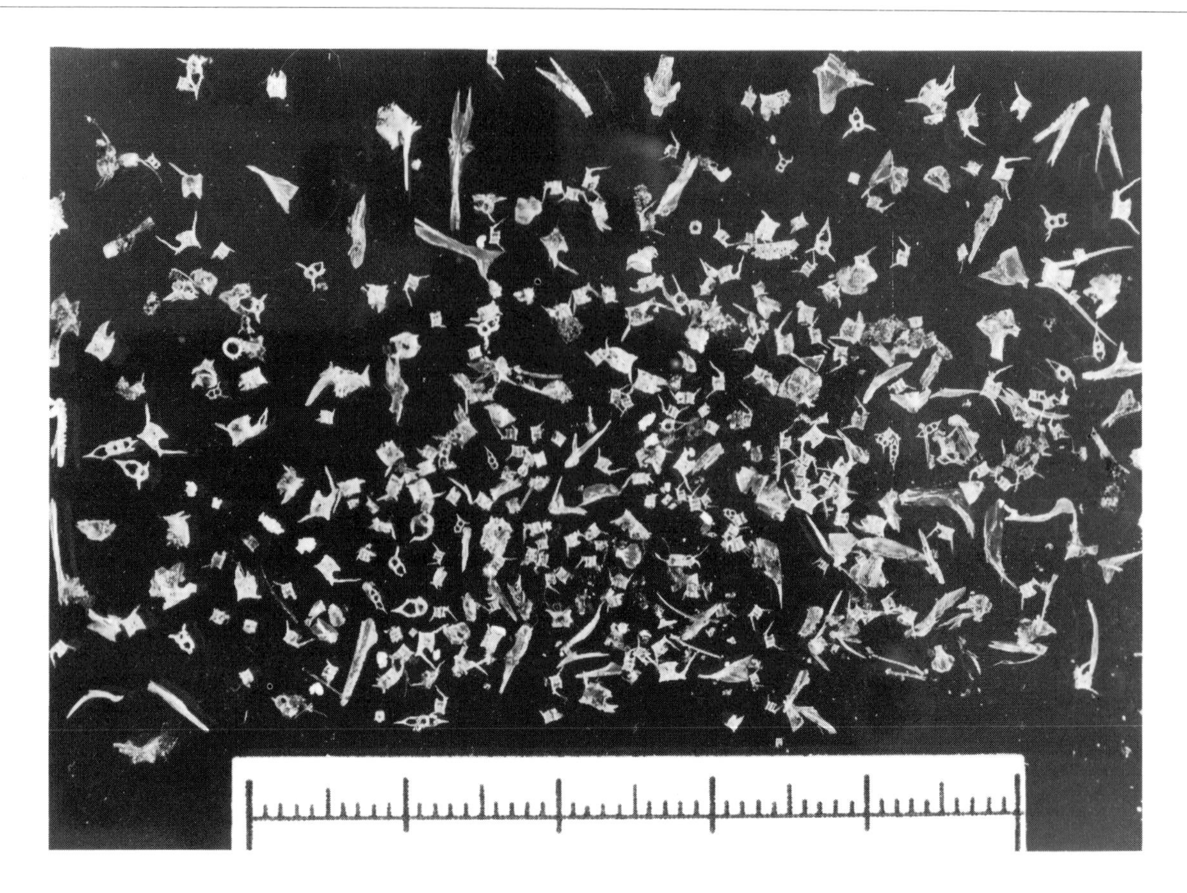

Figure 6.1: Fish bones from a dissagreggated otter spraint.

The similarity in species composition and fish size in all the assemblages, and the exceptional preservation of the small bones both in the modern spraint (Figure 6.1) and in the material recovered archaeologically indicated that a statistical comparison of the skeletal element composition might prove useful. Importantly, as fish species vary in their skeletal structure it is essential only to compare like with like. When spraints from otters living inland were examined (Nicholson 1991) the spraint components were found to be very different from those of the coastal otters, both in prey composition and in the preservation of the bones. A wider prey range was identified in the spraints of the river-dwelling otters, and the skeletal remains were in a much poorer condition than those observed for the coastal otters. Why this should be the case is not clear, but may relate to differences in the size of prey captured (larger fish tended to be represented in the riverine spraints) or to the species of fish captured (some freshwater fish seem to have less robust bones than many marine fish) or to the abundance of food at coastal sites, causing the prey to be less thoroughly digested.

To test the hypothesis that the assemblages of small fish bones from Pool, Tofts Ness and the Links of Noltland were statistically similar in skeletal element composition to modern coastal otter spraint, Spearman's Rank Correlations were calculated for the proportional representation of skeletal elements for each modern spraint group and for the archaeological assemblages (Table 6.2). This non-parametric method was chosen because the data were not normally distributed, necessitating ordinal rather than interval based statistical comparisons. The analyses indicate where the ranking of variables (rather than actual numbers) within a group are more similar to the rankings of the same set of variables in another group than could be expected by chance. Only bones from small fish of less than about 250 mm. were used in the analyses, as in general larger bones were less well preserved and larger fish are unlikely to have been eaten whole, if indeed otters were the agent responsible for transporting the larger bones, which is by no means certain. Otoliths were discarded from the statistical comparison (although identified to species where possible) as they may be lost in preference to bones in some sediments, which invalidates their use in this analysis. Eye lenses were excluded for the same reason. Both were common in the modern spraints.

In both the Freswick and Shetland spraint groups, vertebrae were proportionally the most commonly represented bones, followed by the basioccipital in the case of the Freswick group and the cleithrum in the Shetland assemblage (Table 6.2 and Figure 6.2). Other well-represented bones in both groups included the articular, premaxilla, parasphenoid, dentary and, to a lesser extent, the

Table 6.2: Proportional representation (PR) of skeletal elements from a) modern coastal otter spraints and b) archaeologically recovered accumulations of small fish bones, with Spearman's Rank Correlations.

		Modern coastal spraints				**Archaeological material**									
		Freswick		Shetland		Pool		Tofts Ness		HO86-64		HP90-34		HQ89-31	
MNI =		8		47		18		30		13		11		10	
% vertebrae chewed		15		22		51		41		26		33		37	
% head bones chewed		2		6		7		3		-		-		-	
Skeletal Element	No. in 1 fish	no.	PR	no.	PR	no.	PR	no.	PR	no.	PR	no.	PR	no.	PR
Prevomer	1	7	87.5	38	80.9	6	33.3	9	30.0	4	30.7	1	9.0	2	20.0
Parasphenoid	1	6	75.0	30	64.4	1	5.6	2	6.7	1	7.7	2	18.2	3	30.0
Basioccipital	1	8	100.0	25	53.2	10	55.6	17	56.7	2	15.4	2	18.2	3	30.0
Premaxilla	2	13	81.3	75	79.8	29	80.6	33	55.0	23	88.5	20	90.0	13	65.0
Maxilla	2	9	56.3	68	72.3	16	44.4	30	50.0	2	7.7	7	31.8	4	20.0
Dentary	2	14	87.5	72	76.6	14	38.9	37	61.7	15	57.7	22	100.0	16	80.0
Articular	2	11	68.8	86	91.5	33	91.7	50	83.3	26	100.0	15	68.2	19	95.0
Quadrate	2	4	25.0	65	69.1	23	63.9	34	56.7	19	73.0	7	31.8	3	15.0
Hyomandibular	2	10	62.5	70	74.5	18	50.0	9	15.0	0	0.0	1	4.5	1	5.0
Preopercular	2	7	43.8	53	56.4	17	47.2	13	21.7	10	38.5	11	50.0	6	30.0
Opercular	2	11	68.8	73	77.6	17	47.2	32	53.3	5	19.2	8	36.4	5	25.0
Interopercular	2	11	68.8	42	44.7	9	25.0	12	20.0	0	0.0	1	4.5	0	0.0
Epihyal	2	2	2.5	51	54.3	6	16.7	23	38.3	-	-	-	-	-	-
Ceratohyal	2	7	43.8	79	84.0	17	47.2	14	23.3	6	23.1	5	22.7	3	15.0
Infrapharyngeal	2	8	50.0	53	56.4	7	19.4	8	13.3	2	7.7	0	0.0	0	0.0
Suprapharyngeal	6	6	13.0	16	5.7	7	5.6	29	16.1	-	-	-	-	-	-
Post-temporal	2	3	12.8	12	12.8	7	19.4	8	13.3	0	0.0	0	0.0	1	5.0
Cleithrum	2	7	96.8	91	96.8	13	36.1	9	15.0	0	0.0	3	13.6	4	20.0
Supracleithrum	2	7	42.6	40	42.6	6	16.7	6	10.0	0	0.0	1	4.5	0	0.0
Vertebra	45#	546	151.7	2885	136.4	662	81.7	1032	76.4	1185	202.6	365	73.7	421	93.6
Otolith	2	1	6.3	75	79.8	1	2.8	29	48.3	-	-	-	-	-	-

– 45 is used as an appropriate average, '–' not counted (otoliths from the Links of Noltland had been separated from the rest of the fish bone during post-excavation sorting, and so could not be related to the sub-samples.

Spearman's Rank Corrrelations:
Freswick & Shetland, rho = 0.463 (95%)
Pool & Freswick, rho = 0.440 (94%)
Pool & Shetland, rho = 0.622 (99%)
Pool & Tofts Ness, rho = 0.752 (99%)
HO86–64 & HP90–34, rho = 0.725 (99%)
Tofts & Freswick, rho =0.445 (95%)
Tofts & Shetland, rho = 0.488 (95%)
HO86–64 & Freswick, rho = 0.640 (99%)
HO86–64 & Shetland, rho = 0.708 (99%)
HO86– & HQ89–31, rho = 0.824 (99%)
HP90–34 & Freswick, rho = 0.474 (95%)
HP90–34 & Shetland, rho = 0.527 (95%)
HQ89–31 & Freswick, rho = 0.649 (99%)
HQ89–31 & Shetland, rho = 0.647 (99%)
HP90–43 & HQ89–31, rho = 0.892 (99%)

N.B. To prevent bones being counted more than once, only bones >50% complete have been included. Otoliths have been excluded from the correlations as they may survive differently to bone in some soils.

hyomandibular and opercular. Poorly represented bones included the suprapharyngeal, post-temporal and supracleithrum. In all the archaeological assemblages the most commonly represented bones were the articular, dentary, premaxilla and the vertebrae. In one of the Links of Noltland samples the vertebra was proportionately the most common element: in the other assemblages jaw elements were proportionally more, or similarly, frequent. Unlike the modern groups, the parasphenoid was very poorly represented, as was the prevomer. The cleithrum was poorly represented in the Links of Noltland material. The epihyal, supracleithrum, post-temporal, suprapharyngeal, infrapharyngeal and interopercular were poorly represented in the archaeological assemblages, as in the modern spraint groups. Where wrasse were present, however, their chunky pharyngeal bones were common, as

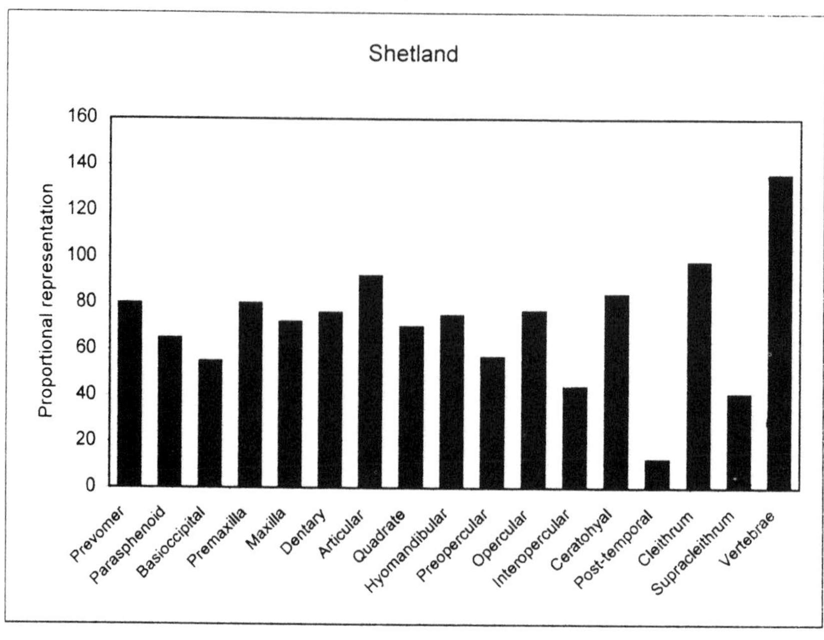

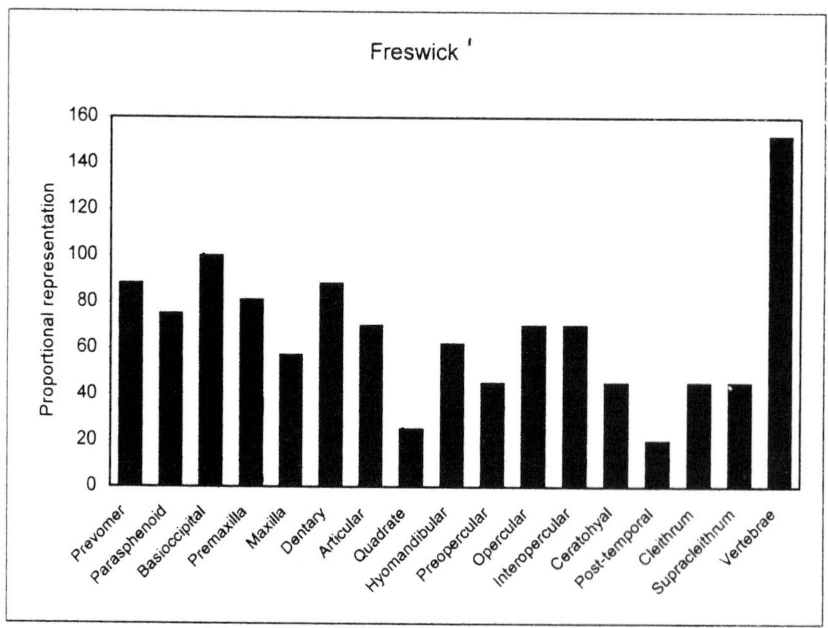

Figure 6.2: Proportional representation of fish skeletal elements from otter spraints collected from a) Shetland and b) Freswick, Caithness.

was the distinctive cottid preopercular. Despite the apparent differences, correlations revealed the skeletal element composition of the Pool and Tofts Ness assemblages to be significantly similar to each other (99% confidence), to the Shetland assemblage (99% and 95% confidence) and to the Freswick assemblage (94% and 95% confidence). All three samples from the Links of Noltland were significantly similar to the modern spraint assemblages (all at 95% confidence or above).

A number of bones from each assemblage was chewed or crushed. Table 6.2 gives the proportions from the contemporary and archaeological groups (the numbers of chewed head bones were not counted for the Links of Noltland samples). In all cases, chewed or crushed bones were more often vertebrae than head bones, and the vertebrae tended to be crushed in the medio-lateral plane, a consequence of which was to break the struts supporting the two articulating facets. This breakage was seldom sufficient to separate the two facets. Larger vertebrae in particular tended to show signs of having been chewed (Figure 6.3), especially on their articulating surface, where puncture marks were often seen. Broken bones were few

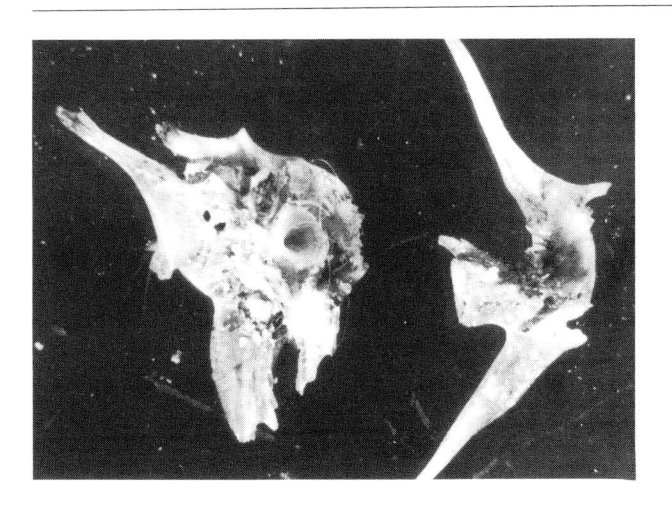

Figure 6.3: Detail of chewed vertebrae from an otter spraint.

in the modern spraint, but the bones most commonly affected were the cleithrum and opercular bones. Table 6.3 gives the proportions of whole bones in each modern spraint assemblage and for the archaeological material from Pool and Tofts Ness (the percentage figure describes the proportion of the whole bone which is represented by the fragment). The archaeological material had evidently suffered greater bone loss and fragmentation than the fish represented in the modern spraint. This may be attributed to the much greater length of time that the archaeologically recovered bone had lain on the ground surface, although the bones did not appear weathered or eroded; in fact some – particularly the smallest bones – appeared almost fresh.

There was no evidence of acid dissolution, either on the bones from the modern spraint or in the archaeological assemblages. Many of the bones from the Links of Noltland samples were coated with a hard calcareous deposit (probably calcium phosphate), similar in appearance to concretions around bones from latrine pits and forming the matrix of mineralised coprolites. This concreted material was seen rarely on bones from the Pool and Tofts Ness samples. A proportion of the bones from the modern spraints were stained black or dark brown, from the tarry mucus which held the spraint together (Figure 6.4). Similar staining was noted on some bones from all the archaeological assemblages. However dark mottling is often seen on bones which have been buried, due to contact with organic acids, iron or manganese in the soil, so this criterion is not considered to be diagnostic of a spraint origin.

Table 6.3: Fragmentation patterns: numbers of fragments in each of four completeness categories, and percentage of whole bones in the modern spraints and archaeological material (fragment completeness describes the proportion, as a percentage, of the whole bone represented by the fragment).

	Fragment Completeness				
	0-30 %	40-60 %	70-90 %	100 %	% whole
Modern					
Shetland	143	215	287	2735	80.9
Freswick	70	57	45	373	68.4
Archaeological					
Pool	365	342	141	482	36.2
Tofts Ness	239	461	148	881	50.9

N.B. The figures for whole bones (100%) include vertebre with only the processes and/or spines missing. Only bones are included; i.e. eyelenses and otoliths are excluded.

4. DISCUSSION

It is inherently unlikely that wind or water action were responsible for the archaeological accumulations of bone discussed here, as they were found inland, away from the strand lines and associated with structures. Additionally, experiments have shown that bones eroded by water or wind action appear smoothed, and vertebrae show characteristically squared edges (Nicholson 1991, 1992).

The fact that most bones are complete in modern coastal otter spraint assemblages poses a problem in trying to interpret assemblages of archaeologically recovered small bones. Fish placed as a "ritual" offering would also appear complete and in good condition if they had been placed in a sheltered situation, away from sun, rain, acidic groundwater and predators. It is known from historical records that small gadids and fish found close to the shore and in rock-pools were fished and consumed on Orkney and Shetland from the sixteenth to the nineteenth centuries (Colley 1983b). Such fish, if later rejected, may also account for deposits of small, well-preserved bones; however in the archaeological circumstances discussed here that interpretation is unlikely. It is more probable that such fish would have been disposed of either into the sea or onto a midden, with other household waste. The archaeological deposits studied were either inside structures, or, in the case of Pool, on a paved area; in none of the cases did they appear to form part of a general midden containing other types of refuse. In any case, if fish were deposited as whole specimens, either as a ritual offering or as discarded fish or fish frames, then the bones should not appear chewed. That some of the archaeologically recovered bones had been eaten is beyond doubt, so the crucial consideration was whether otter spraint could be

Figure 6.4: Detail of fish bones from an otter spraint; note the staining of some bones.

differentiated from the excreta of other species preying on coastal- dwelling fish.

Experiments into the appearance of fish bone after complete fish had been consumed by a dog, pig, rat and human (Jones 1984, 1986; Nicholson 1991, 1993) indicate that in all cases the great majority, or in the case of rat all, of the bones are destroyed. Those bones which survive are very badly corroded and broken. Similarly, seal scats in general contain very corroded bone (Figure 6.5), although small otoliths survive particularly well, and some bones may be in a good condition (Nicholson 1991). Figure 6.6 illustrates the proportional representation of skeletal elements from a sample of 22 grey seal (*Halichoerus grypus*) scats; a similar histogram has been published for the survival of herring bones after consumption by a human (Nicholson 1993). If any of these predators had been responsible for the assemblages, then the bones must have been spat out or vomited for them to exhibit so little damage. Sea birds also prey on small fish, and their pellets, formed of bone and other ingested items which are vomited up, also may account for accumulations of small fish bone. However, studies by the author of the contents of pellets from the Isle of Great Cumbrae, Scotland, almost certainly from either the herring gull (*Larus argentatus*) or the greater black-backed gull (*Larus marinus*), found that the pellets were very diverse, many containing a range of items, in particular crushed shells and crustacean carapaces in addition to fish bone. The gulls also seemed to prey on larger fish than the otters; most bones were from gadids of in excess of 300 mm. The extent of damage to the fish bone found in gull pellets was very variable, but no bone showed features which could be mistaken for the marks caused by chewing using teeth.

With the exception of Tofts Ness, where the bone seems to have been associated with an abandoned building, it is not clear whether the archaeologically recovered bones were deposited during the use or the abandonment of the structures with which they were associated. If otter spraint then the latter is implied, since otters avoid human habitation. In all cases, the evidence favours a spraint origin.

5. CONCLUSIONS

Although it has not proved possible to provide a "law" of formula by which large deposits of fish bone recovered archaeologically can be identified unequivocally as all being from otter spraint, a set of criteria has been developed by which deposits of fish bone from otter spraint may be distinguished from accumulations of bone deposited by wind, water or other predators. By looking in detail not only at the sizes and species of fish represented but also at the extent and types of damage both to the skeletons and

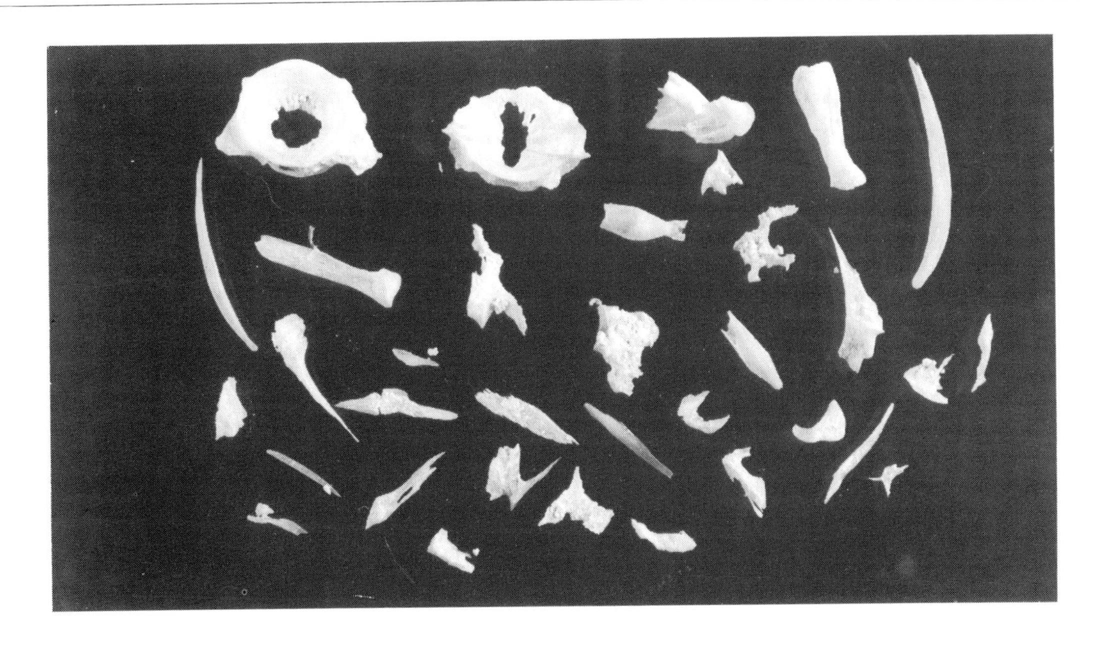

Figure 6.5: Fish bones from a dissagreggated seal scat.

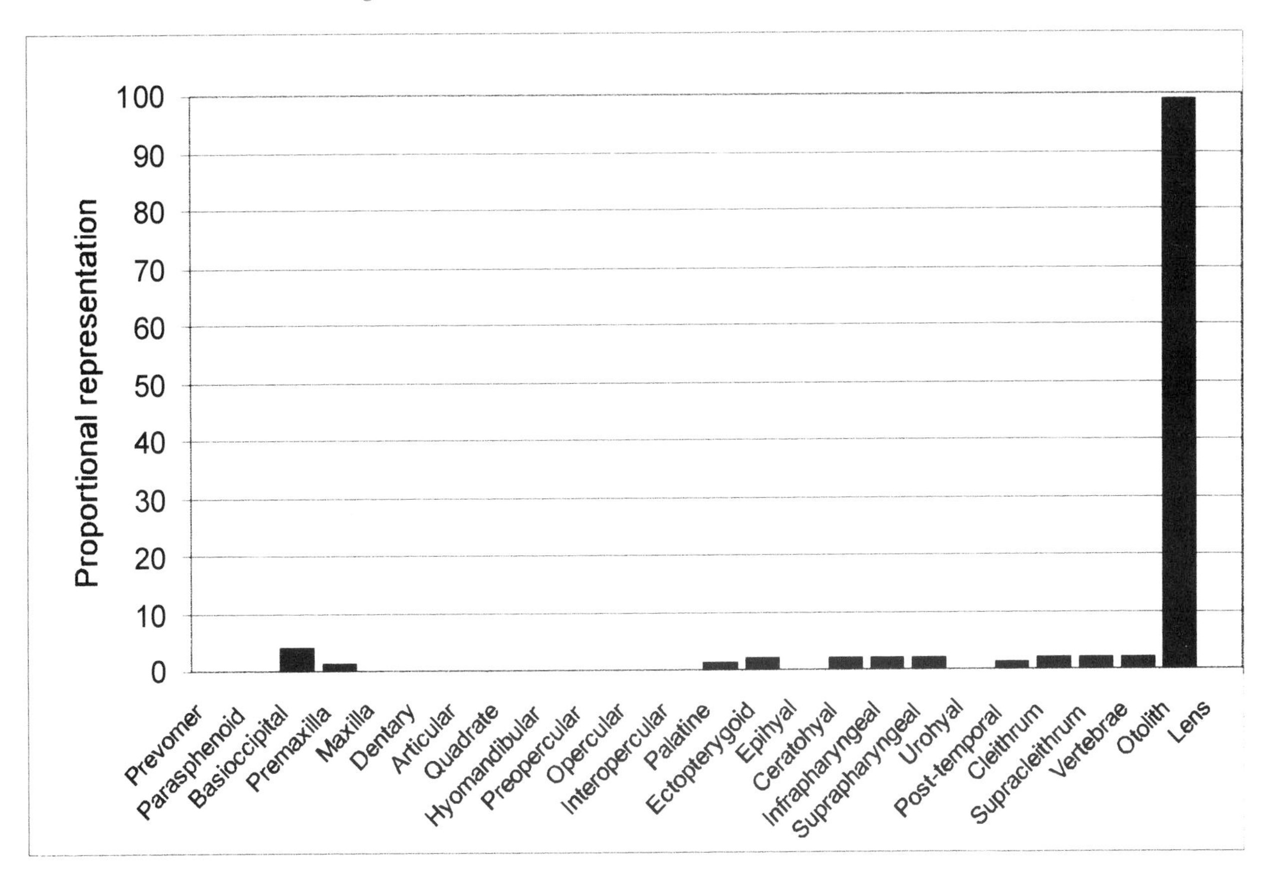

Figure 6.6: Proportional representation of fish skeletal elements from seal scats.

to the individual bones, the presence of otter spraint may be detected even if questions still remain about the origin of some components of the original assemblage. In the archaeological deposits discussed in this paper, the balance of evidence favoured a spraint origin for all, or most, of the recovered material

ACKNOWLEDGEMENTS.

Thanks to Dr. John Hunter, Steve Dockrill and Dr. Julie Bond (Department of Archaeological Sciences, University of Bradford) for supplying the bone, and background information, for the excavations at Pool and Tofts Ness.

Excavations at the Links of Noltland were undertaken by Dr. David Clarke of the National Museums of Scotland, and the assessment work of which a part is discussed in this paper was jointly undertaken with Dr. Andrew Jones (Archaeological Resource Centre, York Archaeological Trust) and forms part of the post-excavation project managed by Alexandra Shepherd (National Museums of Scotland). All excavations were funded by Historic Scotland (formerly the Scottish Development Department), and post-excavation programmes have also been partly funded by the National Museums of Scotland. Thanks also to Andrew Jones, Gordon Woodruff (Department of Biology, University of York) and Dr. Jim Conroy (Institute of Terrestrial Ecology, Banchory) for supplying the modern otter spraints, and to Dr. Graham Pierce (Department of Zoology, University of Aberdeen) for supplying the seal scats.

BIBLIOGRAPHY

Chanin, P. (1987). *Otters*. Mammal Society Series, Anthony Nelson Ltd. Oswestry.

Colley, S.M. (1983a). Chapter 6. The marine faunal remains: In J.W. Hedges (ed.) Isbister, a Chambered Tomb in Orkney. *British Archaeological Reports. British Series* 115, 151–158.

Colley, S.M. (1983b). Interpreting prehistoric fishing strategies: an Orkney case study: In C. Grigson and J. Clutton-Brock (eds.) Animals and Archaeology: 2. Shell Middens, Fishes and Birds. *British Archaeological Reports. International Series* 183, 157–171.

Colley, S. M. (1987). Appendix 1: The marine faunal remains: In J.W. Hedges, B. Bell and B. Smith, Bu, Gurness and the brochs of Orkney. *British Archaeological Reports British Series*, 163–165.

Earlinge, S. (1967). Food habits of the otter, *Lutra lutra* L. in south Swedish habitats. *Vittrevy* 4, 372–443.

Earlinge, S. (1968). Food studies on captive otters *Lutra lutra* L. *Oikos* 19(2), 259–270.

Hunter, J.R., Dockrill, S.J., Bond, J.M. and Smith, A.N. (forthcoming). Investigations in Sanday, Orkney. *Society of Antiquaries of Scotland Monograph.*

Jones, A.K.G. (1984). Some effects of the mammalian digestive system on fish bones. In N. Desse-Berset (ed.) 2emes rencontres d'archaeo-ichthyologie. Centres de recherches archeologiques, notes et monographies techniques, no 16. C.N.R.S. Sophia Antipolis – Valbonne, 61–66.

Jones, A.K.G. (1986). Fish bone survival in the digestive tract of pig, dog and man : some experiments. In D.C. Brinkhuisen and A.T. Clason (eds.) Fish and Archaeology. *British Archaeological Reports International Series* 294, 53–61

Mayson, C.F. and MacDonald, S.M. (1986). *Otters, Ecology and Conservation*. Cambridge University Press. Cambridge.

Nicholson, R.A. (1991). *An investigation into variability within archaeologically recovered assemblages of faunal remains: the influence of pre-depositional taphonomic factors*. D.Phil thesis, University of York.

Nicholson, R.A. (1992). Bone survival: the effects of sedimentary abrasion and trampling on fresh and cooked bone. *International Journal of Osteoarchaeology* 2, 79–90.

Nicholson, R.A. (1993). An investigation into the effects on fish bone of passage through the human gut: some experiments and comparisons with archaeological material. *Circaea* 10(1), 38–51.

Nicholson, R.A. and Jones, A.K.G. (1992). An assessment of the fish bone from excavations at the links of Nottland, Orkney, and recommendations for future investigations. Unpublished report, University of Bradford Department of Archaeological Sciences.

Wheeler, A. (1979). Appendix G. The fish bones: In A.C. Renfrew Investigations in Orkney. Report Res. Comm. Soc. Antiq. London, 38.

7. The butcher, the cook and the archaeologist

P. R. G. Stokes

SUMMARY

An alternative interpretation of the fragmentary long bone waste from Roman and early post Roman sites, from Britain and northern Europe is offered. This project developed from ideas developed for a B.A. dissertation and a discussion with Miss L.J. Gidney, after being shown a Virol jar.

1. INTRODUCTION

On a number of archaeological sites there exist large discrete dumps of bone waste, principally comprising fragmentary long bones. Some of these bone dumps have had preliminary interpretations made by the excavator rather than the archaeozoologist. Other similar dumps have had questions posed by the analyst regarding the nature of the deposit.

For this study five sites were found which had easily accessible reports on the anatomical composition of the bone dumped. These are: Castle Street, Carlisle (Rackham, 1991); York (O'Connor, 1988); Little Chesters (Askew, 1961); Piercebridge (Rackham and Gidney, unpub.) and Zwammerdam (Mensch, 1976). Of these, Carlisle had the largest group, Little Chesters the smallest, while York had a slightly different composition of the elements present. The composition of these five dumps are listed in Table 7.1.

The deposit from Castle Street, Carlisle, (Rackham, 1991) dated to the 8th/9th centuries AD. This exceptionally large deposit was made up of limb bones, including the articular ends, to the almost total exclusion of all other elements. Similar patterns of predominantly limb bones, but with the presence of some other elements, were also found at Little Chesters (Askew, 1961) in a 1st/2nd century AD context; Zwammerdam (Mensch, 1976) in a 2nd/3rd century AD context and Piercebridge (Rackham & Gidney, unpubl) in a 3rd/4th century AD context. The late 2nd century deposit from York (O'Connor, 1988) is remarkable in that it was comprised solely of shaft fragments.

Rackham (1991, 88) suggests that the butcher was boning out the skeleton for some other trader; he goes on to suggest that the limb bones contain supplies of marrow and fats and that they may have been an economic resource in the past. Rackham also points out that marrow can be extracted from other bones, such as metapodials, but which are absent from both Carlisle and Zwammerdam. This pattern is not surprising if these bones represent the waste from secondary butchery. The metapodials were most probably detached from the carcase at the primary slaughter and skinning stage and would, as today, have gone straight to the middleman dealing in hides and offal. To demonstrate this, a calf skin with the feet still attached was collected from a local abattoir and exhibited at the original presentation of this paper. The metapodials would not only travel with the hide, and therefore be unavailable to those collecting long bones from the dressed carcase, but would also be required by the tanner and so would not be accessible at this stage of processing. The tanner utilises the fat content of the metapodials to make Neats foot oil, which is used as a leather dressing on the cured hides (Serjeantson, 1989).

O'Connor (1988, 116–7), in his discussion on the assemblages from the Roman and post Roman deposits from the General Accident Site, York, also suggested that it was difficult to interpret the smashed-up cattle limb bones other than in terms of marrow recovery. For the bone material from Little Chesters, Askew (1961, 108) also suggested the fragmenting of the bone for the extraction of marrow, together with the bones being partly stripped of meat and then being stewed for broth. Another suggestion made to account for this deposit was that the bones were broken up and boiled down for glue.

For the Zwammerdam material Mensch (1974, 163) has a quite different suggestion, stating that the way in

Table 7.1: Site comparisons.

			Sites		
	Little Chesters	York	Zwammerdam	Piercebridge	Castle St., Carlisle
Date (AD)	1st/2nd C	late 2nd C	2nd/3rd C	3rd/4th C	8th/9th C
Element					
Prox Humerus	X		X	X	X
Shaft	X	X		X	X
Dist Humerus	X		X	X	X
Prox Radius	X		X	X	X
Shaft	X	X		X	X
Dist Radius	X		X	X	X
Ulna	X		X	X	X
Prox Femur	X		X	X	X
Shaft	X	X		X	X
Dist Femur	X		X	X	X
Prox Tibia	X		X	X	X
Shaft	X	X		X	X
Dist Tibia	X		X	X	X
Metapodials	X			X	
Other Elements	X		X	X	

which the epiphyses were chopped was characteristic and indicates clearly that the bones were used for extracting broth. Why?

2. THE EXPERIMENTS

The author undertook experimental work to test a theory that there was more to these archaeological deposits than a single product.

Plates XVII, XVIII & XIX from the Zwammerdam report were studied to ascertain how the archaeological bones had been chopped. It appeared to the author that the proximal and distal ends were trimmed after the meat had been stripped off. Mensch, however, suggested that the damage to the proximal and distal ends of the bones had occurred during disarticulation of the carcase. Since giving the original paper, the author has had the opportunity to skin and butcher two cattle of comparable size to Roman beasts. One carcase was suspended for butchery and the other was entirely dismembered on the floor. From this work the author does not believe that such damage would have been incurred during normal carcase dismemberment. To replicate the butchery of the Zwammerdam bones, three complete defleshed long bones, one humerus and two radii and ulnae, were obtained from a local abattoir. It may be noted in this context, *contra* Rackham (1991, 88), that these complete long bones were normal butchery discard and had not been specifically boned out for a secondary user. A wood axe (circa 0.5kg) was borrowed to chop them - this was the only relatively heavy sharp implement that the author could obtain.

The humerus was chopped up in five stages,

1) Distal condyle removed at 45 degree angle to main shaft.
2) Proximal end split following the groove between capitulum and major tuberculum.
3) Capitulum chopped off 90 degrees to shaft.
4) Distal medial condyle chopped off at 90 degrees to shaft
5) Shaft of bone stood on end and split longitudinally to extract marrow

Both ulnae and radii were chopped in three stages:

1) Head of ulna chopped off through shaft below olecranon.
2) Distal end chopped off through shaft above epiphysial line. NB carpals were still attached.
3) Proximal end chopped through groove between articular facets to remove lateral facet. This split the whole shaft sufficiently to obtain the marrow.

Relatively little force was needed in chopping up the bones with the exception of stage two in the radius which required at least three good hard blows with the axe.

The experimentation with these three long bones has shown that, by chopping them in a similar way to those from Zwammerdam, the marrow is extractable without any further processing. If the bones simply were to be turned into stock/soup they would not have needed to be broken up so small, or even broken up at all. Whole bones were, in fact, used by the author in making white stocks when working as a *commis* chef at the Waldorf Hotel in the 1960s. Bones were only sawn into pieces when making brown stock, although this was only done to increase the surface area of the bone prior to being roasted in the oven. The greater the area of browned bone the darker the stock. The fat obtained from boiling up the bones for stock was used, in the recent past, for rouxs (mixture of flour and fat to thicken sauces) and for frying the vegetables to give extra flavour to brown stocks and sauces. The fat that we skim off today would not necessarily have been skimmed off in the past.

Therefore the way that the Zwammerdam bones were butchered does, indeed, strongly suggest that the primary product was the extraction of marrow. Further processing can be postulated but not proven.

To ascertain what further by-products might be obtained, all the fragments of the experimental bones were gathered together, placed into a large pan and simmered for approximately one hour. This was done to release the other fats in the bone (bone grease). The pan was removed from the heat and left to go cold so that the fat could be removed easily from the surface of the liquid. It was thought unlikely that this liquid would have been wasted. It seemed probable that it could have been reduced down to make glue. To test this hypothesis, the bones and the cold liquid were again heated and the liquid was allowed to reduce. The bones were removed when the soft connective tissues had fallen off but before the reduction had become thick and sticky. The liquid was then completely reduced. These processes did not result in any softening or degradation of the experimental bone fragments.

In fact, the current experiment did not produce an edible stock because, before the bones were adequately boiled, they had become somewhat putrescent. It is possible that the processors would have received their bones in the same state, especially during the warmer months and if they came from the butcher and not from the abattoir. Further experiment, after giving the original paper, on timing the butchery of a similar sized beast to that found in Roman Britain confirmed that the bones would almost certainly be sticky by the time the meat had been removed. However, the marrow was unaffected and the secondary fat could cleaned by further processing.

The smashed bone, the bone marrow, secondary fat and the glue obtained from the experiment were weighed. The cleaned humerus, radius and ulna of a Dexter cow, (Woodmagic Meadow Pipit who is part of a reference collection belonging to Ms Louisa Gidney), were also weighed. The Dexter cow radius used for this comparison was of identical size to that of a radius from the Roman Fort at South Shields dating to the late 3rd/early 4th centuries. The weight of the Dexter bones were compared to the weight of the experimental smashed bone and a ratio of the two was calculated. The weights of the by-products were adjusted accordingly to work out the weights of possible by-products that could have been obtained from the estimated number of cattle from Castle St., Carlisle, Rackham (1991). The results are presented in Table 7.2.

Table 7.2: Calculated weights of possible by-products from the Castle St., Carlisle assemblage.

Estimated number of cattle:1500	
Product	**weight (kg)**
Marrow	561
Fat	1009
Glue	336

3. DISCUSSION

Mensch (1974, 164) is convinced that the bone was used to make soup, but gives no clear reasons why, except that many Dutch butchers use the same working-method today. He cites Apicius' *pultes,* which he describes as a simple porridge, but does not explain why. It may be that in all the translations the word *ius* is translated as stock in the soup section only. *Ius* can also be translated as soup, broth or sauce. The term *Jus* is still in use today in culinary French for meat gravy. It is also the root word for juice. Nowhere in Apicius is there any reference to stock made from bones nor meat essence (reduced stock).

Making stock with bones is a fairly recent practice in the upper and middle classes. In the latter half of the eighteenth century it was conceived by those who were concerned at the plight of the poor and needy that all their troubles would vanish if only they would learn to make hot, nourishing soups. However, most of them were so poor that they could not even afford fuel let alone food and thus very rarely ate warm or hot food. By the end of the century, charity soup-kitchens were introduced and over the next half century soups/gruels and pauperism were closely linked by the poor.

In 1772 Captain Cook took "portable soup" with him on his voyage around the world, for feeding to sick sailors. Sir John Pringle (1776) in his "A Discourse upon some late Improvements of the Means of Preserving the Health of Mariners", states how the soup was made:

> *"having by long boiling evaporated the most putrescent parts of the meat [it] is reduced to the consistence of a glue, which in effect it is, and will like other glues, in a dry place, keep sound for years together."*

A part of one of Captain Cook's portable soup cakes was analysed 160 years later and was found not to have undergone any marked change (Drummond & Wilbraham, 1958, 315).

A more luxurious portable soup recipe can be found in Elizabeth Raffald's book "The Experienced English House-keeper" which was first published in 1769. This soup was made with three large legs of veal, one of beef and the lean part of a half ham. The meat was cut into small pieces and placed in a large cauldron with a quarter pound of butter and the bones. The soup was flavoured with anchovies, mace, celery and carrots. These ingredients were sweated over a moderate fire and the gravy removed as it ran from the meat. It is of note that water was not added to the meat until all this gravy had been removed. The cauldron was then set to boil gently for four hours. This liquor was reduced, the gravy added and the whole further reduced until it looked thick like glue. At this stage it was poured into earthenware dishes, a quarter of an inch thick. When set it was cut into rounds the size of a crown, dried and stored in a tin with writing paper in between each cake. One cake was used to make one pint of broth.

It can be seen that the bones were very incidental to the manufacture of this kind of soup. This particular recipe does not even explicitly state that the bones should be broken. However the following recipe in her book, for transparent soup, clearly states that the meat should be cut from a leg of veal and the bone broken into small pieces.

Hannah Glasse also gives a recipe for portable soup in her "Art of Cookery Made Plain and Easy" written in 1747 in which she uses two legs of beef and a number of herbs. The bones are not used in the making of the stock.

Mrs Beeton's (1861, 93) portable soup is made from veal knuckles and shin of beef and her first instruction is to remove the marrow from the bones.

Portable soup therefore appears to have been a mid eighteenth century phenomenon and could be seen as the ancestor of the modern stock cube. An exhaustive search of the standard medieval texts on cookery failed to reveal any medieval predecessor for this concoction.

The bone waste from the 18th century process would leave a similar composition of fragmented elements to that found in the Roman deposits under discussion. It is conceivable that the Romans could have been producing a comestible similar to portable soup, particularly as part of the soldier's iron rations. However there is no literary evidence whatsoever to support such a hypothesis. Only bacon, cheese, hard tack (corn) and sour wine appear to be mentioned in surviving texts (Davies 1971, 124). It seems unlikely that the writers would have omitted a product such as portable soup if it was being mass produced.

The late nineteenth century workhouses and prisons were places where gruels and soups were still commonplace foods:

> "*The soup to be made with ox-heads, in lieu of other Meat, in the proportion of 1 ox-head for about 100 Male Prisoners, and the same for about 120 Female Prisoners, and to be thickened with Vegetables and Pease, or Barley, either weekly or daily as may be found most convenient.*" (Drummond & Wilbraham 1957, 367).

It is not until the early 20th century that we see the middle and upper class cookery book advocating the use of bones for stock. Cassell's New Dictionary of Cookery (1912, 866) states:

> "*A few years ago it was customary for cooks to make stock with fresh meat only.....Altered prices have necessitated the adoption of more economical methods, and now excellent stock is constantly made with the bones and trimmings of meat and poultry.*"

All of the recipes for stock in "A Guide to Modern Cookery" (Escoffier 1907) give meat as a base. The effects of two World Wars, and the necessity to make a profit in the hotel and catering industry, possibly gave rise to the widespread use of bones for stock. This practice is now rapidly disappearing due to the widespread use of stock cubes, which are only a crumble version of the portable soup. The bone waste from stock-cube production is now used for bonemeal and so lost to the archaeological record. Glue produced in a similar way is a far more likely explanation for these deposits of fragmented bone because bone glue was until recently one of the best and most readily available source of glue (Hodges 1989, 168).

Mensch also quotes Varro who describes how to feed sheep dogs. Again it is the translation that may be at fault. For example the version of Varro translated by Ash (1933, 401) reads "They are also fed on bone soup and even broken bones". But this could also be translated as "They are also fed on sauce out of bones and even broken bones" since the Latin *ius ex ossibus* can also be translated as the juice out of bones. It is the author's belief that bones were not cooked, merely cracked open and the marrow, which in a warm climate would be semi liquid, spread on the pieces of bone. What shepherd would go to all the lengths involved in cooking soup for a dog? This would require an extra cooking vessel and, more importantly, would consume fire wood, a scarce commodity. Experimentation was again used to see whether or not this theory was feasible. A marrow bone was kept at 30°C all day and then smashed on a large stone with a smaller one. The marrow was a soft semi-liquid, as expected. This was then fed to two dogs, the smaller of the two licked the larger of the two stones clean whilst the larger dog ran off with the bone.

Therefore there is no conclusive evidence for soup making from bones in either Apicius or any other piece of classical literature.

So what was going on at Zwammerdam and elsewhere? Bone marrow extraction must be part of the answer. But what use was or is bone marrow put to? Until the 19th century nearly all the recipes containing bone marrow that the author can find are sweet/pudding recipes. The exception is baked marrow bones served to the Lords of the Star Chamber on nearly every 'meat' day from 1519, when they cost 12d., 16d. and 18d, to 1639 when their cost had risen to 3s. 6d. (Simon, 1952). Baked marrow bone was a popular dish in Georgian times. The bones were sawn and the open top was covered in paste of flour and water to prevent evaporation. They were served wrapped in a napkin with long silver marrow spoon, a practice that the author can remember still being carried out at the Waldorf

Hotel in 1968. In Victorian England this bone fare was served for men, being considered "unladylike", but Queen Victoria ate marrow toast for tea every day (Hartley, 1954). Also at this time marrow was being used as a garnish to some meat dishes.

It is, however, the sweet recipes that give the link back to the Virol jar that started this line of investigation. "Virol" was a bone marrow product, originally formulated by an Australian chemist, Shepperson. As the product was similar to Bovril, Bovril Ltd. bought the formula in about 1877, to protect themselves. Virol was first produced at Bovril's premises in Old Street in London, an area which had a large number of manufacturers producing Extracts of Meat. What volume of bone waste was produced by the manufacture of Virol is unclear as Bovril did not process the bone itself and the company today have no records of their suppliers from that period. The Virol factory moved to Hanger Lane in Ealing in 1921 and it was this move which left us with a permanent record of what Virol was. To mark the move, Virol was packed in earthenware jars, supplied by Doultons from 1921 to 1935.

The original ingredients for Virol were:- Beef fat; Marrow fat (discontinued during the War, substituted by body fat); Red bone marrow solution (discontinued in 1946); Syrup of lemon; Sugar; Fresh eggs (Dried egg used from 1909); Salt; Oil of pimento; Oil of hops; Quinine sulphate; Calcium bisulphate solution; Calcium chloride; Malt extract.

Virol was termed as a food supplement and was given to babies after a feed on a spoon, or stirred into a bottle feed. Children and adults took it from a spoon twice daily (CPC (United Kingdom) Ltd. 1991) (*wonderful stuff – editor's comment*). So is it inconceivable that bone marrow has been used through the centuries as a sweet or a sweetened food supplement? If mixed with either honey or sugar it would keep for some considerable time. Hartley (1954, 90) writes

> *"The only people I have found who appreciate the value of marrow bones are some children's nurses (specially trained in dietetics), and Army cooks who can count on getting a couple of pounds of 'butter' out of the canteen marrow bones".*

Fat is one of the best sources of energy, approximately 9 kcals per gram, it is also an important source of vitamins A & D.

However, there is still no hard and fast evidence, all that we can say is that animal fats have probably been used as a food. What other uses can animal fats be put to?

The possibilities include use as a lubricant or as a skin/ leather dressing. A Mesopotamian tanning formula for goat skin uses fat of a pure cow (Forbes, 1966). Fats have also been used for making soap and cosmetics. Ox fat is mentioned in some very early texts in the making of pre-classical cosmetics (Forbes, 1965). One of the cited texts for cosmetics was a herbal written during the first century A.D. by a Greek physician, Dioscorides of Anazarba, in Cilicia. In The Greek Herbal, (Gunther 1934) Book II there is a recipe for Marrows, which lists in order the animals to be used – calves and bulls are second and third ranked. Dioscorides states that autumn is the best time of year for the marrow. He then goes on to describe the process of cleaning the fat and storage in an earthenware vessel. Finally he tells us how to use it, suggesting that it can be used in a similar way to poultry and goose fat, that being to take away weariness. This again compares with the purported properties of Virol.

Dioscorides' herbal could have been known to those who produced the bone waste at Carlisle. In an early medieval text: Leechdoms, Wortcunning & Starcraft, there is evidence for the use of marrow and bone grease among the ingredients for a salve. In the translation by Cockayne (1865 vol. III, 15), recipe number 12 states

> "then let one collect together all the bones, which can be gathered, and beat the bones with an iron axe, seethe and skim off the grease"

Although the text does not specify that the marrow was removed before boiling, it is implied. Further on in the same recipe, marrow of several kinds is used.

Therefore my personal interpretation of the fragmented bone waste is that it represents the waste from a complex industry producing marrow, marrow products, fats and possibly products such as leather dressing, cosmetics and soap and not just a simple "soup kitchen"..

BIBLIOGRAPHY

Askew, S. (1961) An Excavation on the Roman Site at Little Chester. *Derbyshire Archaeological Journal* 81, 107–8.

Beeton, I. (1861) *Book of Household Management* (facsimile edition 1982) Chancellor Press, London.

Cassell's New Dictionary of Cookery (1912). Cassell & Co., London.

Cockayne, O. (1865) *Leechdoms, Wortcunning & Starcraft of Early England* V3. Longman, Roberts & Green. London.

CPC (United Kingdom) Ltd (1991) Ambrosia Research & Development Dept. Archive Bray, P.A. (archive).

Davies, R. W. (1971) The Roman Military Diet. *Britannia,* 2, 122–42.

Drummond, J.C. & Wilbraham, A. (1957) *The Englishman's Food.* Jonathan Cape Ltd. London.

Escoffier, G.A. (1968) *A Guide to Modern Cookery.* 2nd Ed. 5th impression. Heinemann. London.

Flower, B. & Rosenbaum, E (1958) *The Roman Cookery Book: A Critical Translation of the Art of Cookery by Apicius.* Harrap & Co. London.

Forbes, R.J. (1965) *Studies in Ancient Technology* vol. III. E.J. Brill. Leiden.

Forbes, R.J. (1966) *Studies in Ancient Technology* vol. V. E.J. Brill. Leiden.

Glasse, H. (1747) *The Art of Cookery Made Plain and Easy* (1983 Facsimile edition) Prospect Books. London.

Gerrard, F. (1945) *Meat Technology.* 1st Edition. Hill. London.

Gunther, R.T. (1934) *The Greek Herbal of Dioscorides.* Oxford University Press. Oxford.

Hartley, D. (1985) *Food in England.* Futura Publications. London.

Hodges, H (1989) *Artifacts.* Duckworth & Co. Ltd., London.

Ministry of Agriculture, Fisheries & Food (1990) *Manual of Nutrition. Reference Book 342.* HMSO, London.

O'Connor, T.P. (1988) The Animal Bones from the General Accident Site, Tanner Row In: *The Archaeology of York The Animal Bones 15/2.* York.

Rackham, D.J. (1991) The animal bone from post – Roman context pp 85–88 In: McCarthy, M.R. *The Structural Sequence & Environmental Remains from Castle Street, Carlisle. Excavations 1981 – 2* Fascicule 1. Cumberland and Westmorland Archaeological and Architectural Society Research Series No. 5. Kendal.

Raffald, E. (1782) *The Experienced English Housekeeper.* (1970 Facsimile from 8th edition) E & W Books (Publishers) Ltd., London

Serjeantson, D. (1989) Animal Remains & the Tanning Trade pp. 129–142. In Serjeantson, D. & Waldron, T. (Eds.): *Diet & Crafts in Towns.* BAR British Series No. 199, Oxford.

Sinclair, R. (1964) *Soap-Making.* Curwen Press, London.

Singer, C. (1927) The Herbals in Antiquity *The Journal of Hellenic Studies* 47(1), 1–52 Macmillan & Co., London.

Sweet (1959) *Anglo-Saxon Reader in Prose & Verse.* 14th Ed Clarendon Press, Oxford.

Thorpe, B. (1846) *Analecta Anglo – Saxonica.* New Ed. Smith, London.

Tooley, P. (1971) *Fats, Oils & Waxes.* J. Murray, London.

Van Mensch, P.J. (1979) A Roman soup-kitchen at Zwammerdam? *Berichten van de Rijksdienst Oudheidkundig voor het Oudheidkundig Bodemonderzoek* 24. 159–65

Varro, Marcus Terrentius *De Re Rustica.* ed. Hooper, W.D. & Ash, H.B. (1954) The Loeb Classical Library, London.

Vehling, J.D. (1936) *Apicius, Cookery & Dining in Imperial Rome.* Limited Edition. Chicago.

Wheeler, R.E.M. & T.V. (1936) Verulamium. A Belgic and two Roman Cities Report XI *Society of Antiquaries of London.*

8. Detecting the nature of materials on farms from Coleoptera: a number of taphonomic problems

David N. Smith

1. INTRODUCTION

This study uses modern analogue faunas to examine some of the possible depositional and taphonomic effects that can shape an insect fauna during its formation. Many of these have been considered by archaeo-entomologists for some time (for example the effects of allochthonus species on a fauna's interpretation (Kenward 1975, 1978)). This study builds on such work and presents a recorded example for the archaeo-entomologist to consider and consult.

2. THE NATURE OF THE STUDY

One situation where it has been suggested that a detailed interpretation of archaeological deposits from their beetle faunas could occur is on small farms. In particular, it has been suggested that as fodder and animal bedding move through the different stages of the farming round they should develop distinct and interpretable faunas (Buckland, Sadler and Smith 1994). One possible model for this, based upon a Norse farm, is presented in Figure 8.1. It should be

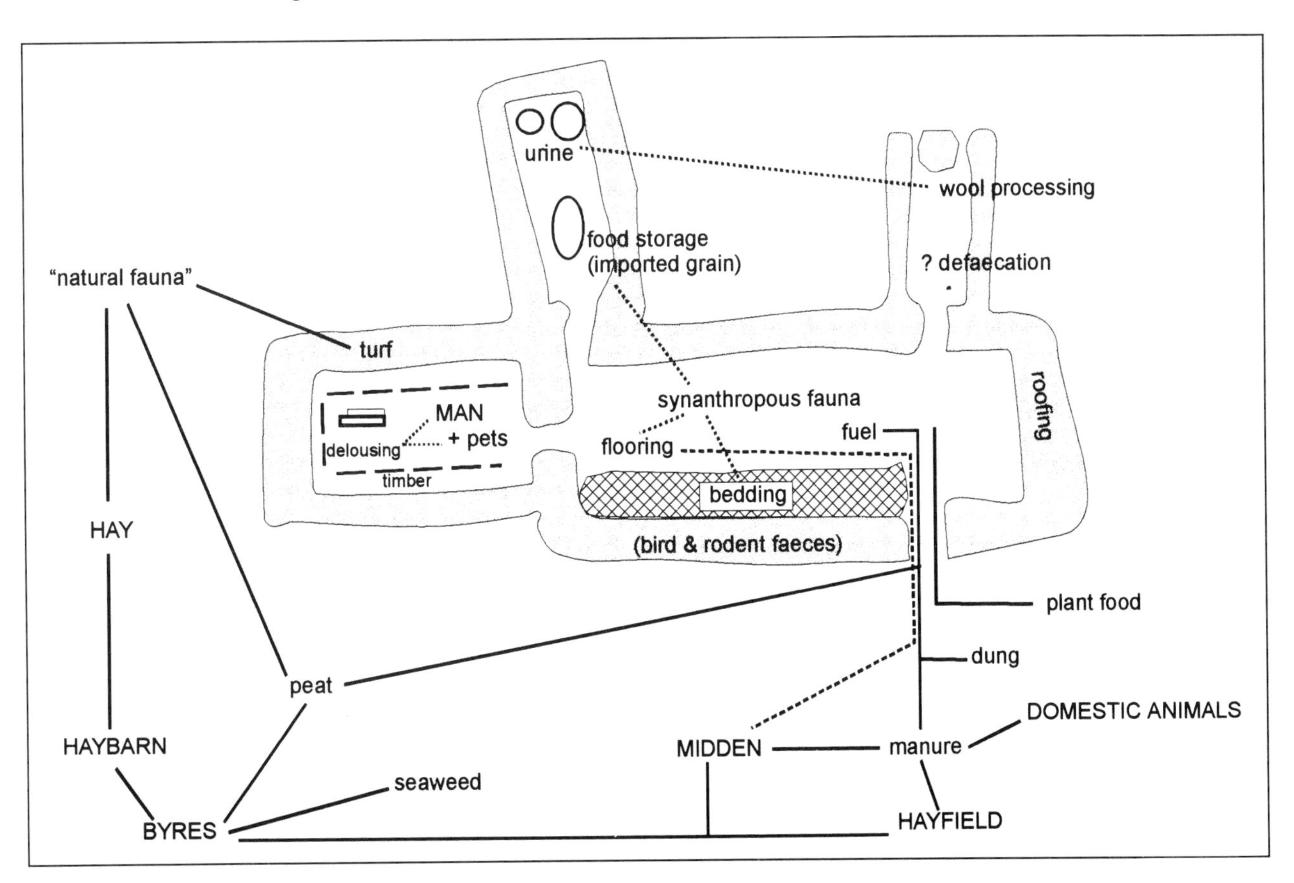

Figure 8.1: The proposed movement of insects and materials around a Norse farm. (redrawn from Buckland, Sadler and Smith, 1993).

remembered that this paper only addresses the rôle of the beetle fauna in this suggested model. The suggested inferences that could be drawn from the presence of Diptera and ectoparasites of humans and animals are not addressed by the study presented here.

However, one problem for this proposed scheme is there are no modern studies from similar circumstances. This model had not been "field tested" to see if the proposed detail is possible. It was therefore decided to survey the differing Coleoptera faunas from a modern simple farming system in order to see how easy to "fingerprint" they are and to examine potential taphonomic problems.

3. SAMPLE COLLECTION

All samples were collected between December 1988 and November 1989 from Conisbrough Park Farms, Conisbrough, South Yorkshire, England. This farm was chosen for two reasons. Firstly, there was no improvement of the pasture nor was insecticide used. Secondly, the practice of permanent deep littering – allowing a bed of straw, hay and dung to build up over winter – was still continuing at this farm. In this stock rearing system 1–2 m beds of hay, straw and dung are allowed to gather all winter and spring after which the byre is cleared and the material is thrown onto the arable fields.

In total 6 types of material were examined. The relationship between these materials and how they move around the farm is illustrated in the flow diagram in Figure 8.2.

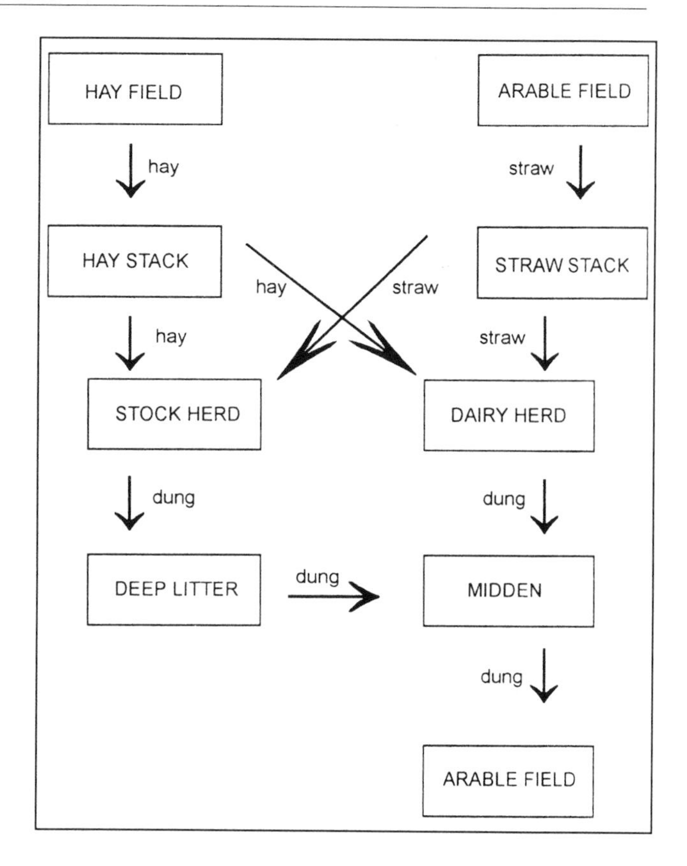

Figure 8.2: The movement of fodder and bedding materials around Conisbrough Parks Farm.

(1) Hay: the collected stalks and heads of grasses taken from the farm's pasture and meadows in the summer of 1988. This material was stored in open sided Dutch barns and used mainly as cattle feed. The material was dry and dusty to the touch.

(2) Straw: the dried stalks from the farm's arable crops. This was kept in the open in large uncovered stacks. The material was dry and dusty. This material was used as cattle bedding.

(3)The wind battered and blackened tops of the straw stacks: this material was often saturated with water and rather cold to the touch.

(4)The deep litter bed: the material that built-up under the stock herd during the winter. All of the samples came from a building with stone walls and a flag floor. This building had very restricted drainage. During the course of winter the deep litter bed reached a depth of 1.5–2m. The bed itself was made up of straw with some hay and was full of cattle manure. The material itself was very compacted and could not be cut with a spade. There appeared to be no "heating" as a result of biological breakdown within the litter. The deep litter was saturated with urine and smelt strongly of ammonia. Samples came from three depths within the bed:

a) the fresh material at the surface (DL1),
b) the compacted, saturated material from the middle depths (DL2) and
c) the waterlogged material resting above the flags of the floor (DL3).

(5) The farm yard midden: this lay in the corner of a large arable field. It consisted of bedding and dung from the dairy herd. Two sample types were taken. "Active midden" was taken from the areas of "heating" within the midden. This material was very hot to the touch. The material was often blackened, dry and covered in a thick white mould. The old midden material had been on the edge of the pile for several months. It was often cold, crusty and dry. Little mould was evident.

(6)Yard compost: a pile of straw and dung accidentally dumped from the deep litter beds over the course of several years. The degree of organic breakdown was advanced and there was a distinctly earthy feel to this deposit. This material dried out considerably during the summer.

SAMPLE PROCESSING

The samples were placed in a cold store running at -6°C until processed. Approximately six litres of material from each sample was measured by packing it loosely into a graduated bucket. The drier material such as the hay and the straw was left overnight in an oven at 40°C. In the morning the material was sieved using 4mm, 2mm and 300µ mesh sieves. Each of the resulting residues was sorted by hand under the microscope and the insect remains collected.

More earthy or faeces ridden material was processed using the standard methods of paraffin flotation outlined in Coope and Osborne (1968) and expanded upon in Kenward *et al.* (1980). The resulting paraffin flots, however, were often very large and had to be dried in an oven overnight and then sieved using a 2mm and a 300µ mesh and the insect remains removed.

The resulting coleopterous material was identified where possible by comparison to the collection housed in the Department of Archaeology, University of Sheffield.

4. RESULTS

The species lists for six of these types of materials are presented in Tables 8.1 to 8.6. The taxonomy follows Lucht (1987) and the numbers in brackets are the numbers of individuals found articulated. Data from the straw samples are not included here due to the low numbers of insects present. Full species lists are, however, included in Smith (1991).

Dead or Alive?

Before considering the entomological nature of these materials it is important to examine if the samples here have produced true death faunas. A major problem for the archaeo-entomologist is that modern faunas often collected as live individuals may not replicate the death assemblage that would be found in a comparable archaeological deposit. Principally, this is the result of the fact that some species may have emigrated from this material had not the investigator intervened. In addition, death assemblages often contain accidental captures and allochthonus elements that would not be present in the live fauna (Kenward 1975, 1978).

There are a number of good reasons to suggest that the majority of the materials examined here do, in fact, represent true death assemblages, or at least good approximations to them.

Firstly, the materials containing the insects were processed and sampled as though they were archaeological deposits. This obviates problems over the use of differing collection and sampling methods.

Secondly, the majority of the insects examined here appear to have been dead at the time of collection since they were disarticulated. This is particularly true of the yard compost, the middens and the deep litter. The results from the straw and hay samples present more of a problem. In these samples many individuals were whole and articulated. It is therefore possible that many of these individuals were collected live. However, for a number of reasons, it would seem that although we are not dealing with a true death assemblage from these materials we may still be dealing with a close approximation to a death assemblage. One reason is that when the articulated individuals are removed from the species counts the occurrence and relative numbers of the species present remains similar. Another reason is that there is some circumstantial evidence to suggest that there is little immigration or emigration of species in these types of deposit. This was suggested by a shaking and pitfall survey which was undertaken concurrently with the main period of sampling. These surveys produced the same species and proportions as the putative death fauna (see Smith 1991).

The entomological nature of the materials examined

(1) The hay

The large faunas in the hay samples are dominated by species commonly associated with mouldering but dry plant materials. This type of material has been described as "sweet compost" by Kenward (1982). The majority of the fauna consists of two species:- *Typhaea stercorea* (L.) and *Cartodere ruficollis* (Marsh.). Other species of importance in this fauna are *Xylodromus concinnus* (Marsh.), the (suggested) modern import *Ahasversus advena* (Waltl.) and the *Cryptophagus* and *Enicmus* species. These species are all predominantly fungal feeders.

T. stercorea is recorded as a common inhabitant of dry mouldering materials in hay lofts, stacks and damp matter in food product stores (Le Clercq 1946, Green 1953, Coombes and Freeman 1956, Hunter *et al.* 1973). The results of the shaking survey from this material suggest that it prefers the slightly damper material around the base and outer edges of the stacks. From this survey it would also seem that *C. ruficollis* prefers the driest material in the centre of the stack (Smith 1991). It would seem *X. concinnus* and the *Cryptophagus* and *Enicmus* species prefer much more damp materials. In the shaking survey they dominated in the rain washed materials at the base of the straw stacks (Smith 1991).

Some of these species have been used to suggest the presence of cut and stored hay in structures and midden layers on a range of archaeological sites (For example Buckland *et al.* 1983, McGovern *et al.* 1983, Perry *et al.* 1985 but see Smith (1998)).

(2) The straw

The samples from the straw stacks contained no or only a few individuals. Where present they were mainly *Cryptophagus, Enicmus minutus* (group) or *T. stercorea*. The explanation for the absence of species from this material is that straw is harvested as field-dried, dead stalks whereas hay is collected as live grasses. Hay will therefore contain a higher nutrient value that would encourage fungal hyphae which, subsequently, encourages a beetle population to develop.

(3) The material from the top of the straw stacks
The material from the top of the straw stacks contained a very distinctive fauna. There are few of the classic members of the "hay" or "sweet compost" fauna seen above. The fauna is dominated by *Micropeplus staphylinoides* (Steph.) and *Acrotrichus* species. It is probable therefore that these species favour extremely exposed and wet materials such as this.

(4) Deep litter
The nine samples from the various levels of the deep litter beds produced a very sparse fauna (not just of Coleoptera but also of Diptera). This small assemblage is made up of members of the "hay" or "sweet compost" fauna. This was not what was expected. It was thought that this mixture of wet plant remains and animal dung would develop a large fauna of species associated with wet and foul compost. One possible explanation for this is that the ammoniac nature of this material prevents the development of the expected insect fauna. It is probable that the members of the "hay" fauna came in either live or as dead sub-fossils when hay was added to the stall. That this material could have supported an insect fauna under other circumstances can be demonstrated. In the summer after the barn had been emptied of cattle and mucked out, and the few inches of material that remained had drained and lost their ammonic smell, a fauna of foul matter loving species developed. Hand shaking of this material produced numbers of *Quedius, Philonthus* and *Anthicus* species.

(5) The midden deposits
The midden deposits produced very different faunas to those seen above. With the old midden many of the species present are commonly associated with very rotten, cold and foul plant matter. Examples of these are the *Cercyon* species, the *Anthicus* species, *X. concinnus* and the *Oxytelus* species. Other species such as the *Quedius* and *Philonthus* species also live in wet and rotting matter. In addition the *Micropeplus* and *Acrotrichis* species, which were seen in the tops of the straw stacks, were also present in this exposed material.

The active midden, however, appears to have contained two contrasting beetle assemblages. Many of the species which favour wet, cold and foul matter are again seen here as they were in the old midden. However, there are also large numbers of individuals from dry and mildly mouldering materials, again "sweet compost". In fact these are the same species which dominated in the hay stacks, namely *D. ruficollis* and *T. stercorea*. This suggests that within this area of the midden two micro-habitats were present. There are areas where the material has dried and mouldered through biological activity or "heating" and others where the material remains cold, wet and less biologically active. When sampling occurred, materials from both of these spaces were combined within the one bag. It seems unlikely that this number of individuals could have been thrown onto the midden in material from the dairy herd particularly as samples of material straight from the dairy herd deposits contained no or few species (Smith 1991).

(6) The yard compost
The yard compost produced a wide ranging fauna suggesting that its constituent materials were from widely differing provenances. In terms of the species present there are elements from many of the other faunas from Conisbrough Farms. Again species from very dry plant materials, such as *C. ruficollis*, *T. stercorea* and the Lathridiidae, were present probably because this dump of material dried out over the summer allowing them to invade. Also present were species from wetter plant materials such as the *Cercyon*, *Omalium*, *Oxytelus* species and the *X. concinnus*. Very rotten and foul materials are again represented by *Cercyon haemorrhoidalis* (F.) and the *Anthicus* species.

One difference when these faunas are compared with the other samples from Conisbrough Parks Farm is that a major part is made up of outdoor species (*sensu* Kenward 1978). These are all species which can breed away from human habitation. This group is made up from an assortment of individuals of various Carabidae, Scarabaeidae, Coccinellidae and Curculionidae. This reflects the relatively open nature of this deposit.

5. THE IMPLICATIONS OF THESE FAUNAS

The faunas seen here suggest that the application of the model presented in Buckland *et al.* (1994), in terms of the Coleoptera, may not be as straightforward as hoped. Many of the Conisbrough Park Farms materials, and the situations from which they were sampled, failed to produce clear "fingerprint" faunas. Many of the faunas were too similar in terms of the type and proportions of species recovered to be readily interpretable if they occurred in the archaeological record. This would suggest that attempts to determine the past of history of archaeological deposits may meet with interpretational problems.

The inability to differentiate these materials by their beetle faunas seems to be due to several very specific depositional problems.

The most obvious message to come out of this study is that there are stages in the movement of this type of material around the farming cycle that are effectively invisible in terms of the Coleoptera present. These stages often produce small or no faunas at all. Two good examples of this are the majority of the straw samples and the deep litter beds. With the straw this is the result of the nature of the materials themselves. However, with the deep litter beds the nature of the fauna is dictated by an external factor. In this case, the blocked drainage within the barn resulting in the pooling of ammoniacal urine. This situation could not be predicted or detected by the archaeo-entomologist from the beetle fauna alone. One possible result of this is that a deposit with a similar fauna would probably be interpreted as a building containing "sweet compost" or hay if it was found in the archaeological record.

The second taphonomic effect demon-strated by this study is that as materials pass through the farming system the remains of dead individuals from the preceding stage can get carried along. The prime example of this is the species from the "sweet compost" or "hay" fauna that occurred in the damp and foul deep litter bed. On one hand this effect acts to blur the interpretation of a deposit's nature but under some circumstances could allow some assessment of its past use.

A third effect seen in this material is a problem generated by the nature of bulk and archaeological sampling. With some of the materials seen here, such as the active midden and the yard compost material, we have insect assemblages from two differing micro-habitats which were combined when sampled. It is possible that a similar fauna from an archaeological midden could be misinterpreted. For example, one possible interpretation is that the two contrasting ecologies result from the combining of two distinct materials from two sources on the farm. The dry matter, for example, may have come from hay stores and the fouler matter from decayed flooring or from stable materials. However, this is clearly not the depositional situation with the modern midden at Conisbrough Park farms. Here we have cold, damp matter which has dried and subsequently developed a "dry hay fauna" within the midden itself. This suggests that the inference of the presence of "hay" or stable deposits just from the coleopterous members of the hay fauna alone is dangerous. This is a point expanded upon in Smith (1998).

The presence of hay and stable manure is probably better indicated by the occurrence of this fauna in combination with the presence of phytophage beetles from grasslands and pastures such as *Apion, Sitona, Hypera* and *Gymetron* species. In addition, supporting evidence for cut hay should be gained from the plant macro and pollen evidence. Stable manure is probably best identified by the use of Diptera, lice and flea remains. This approach has been followed on a number of archaeological sites (Grieg *et al.* 1982, Hall and Kenward 1990) and discussed extensively in Kenward and Hall 1997, Hall and Kenward 1998.

All of the three factors here outlined could serve to blur the palaeoentomological "signature" for these materials and result in a substantial information loss. The relative importance of each factor would be difficult to assess when dealing solely with fossil data.

6. CONCLUSIONS

The taphonomic effects described in this study suggest that identifying the materials examined here by their Coleoptera faunas alone would be difficult. In the main the materials and their faunas are just too similar, and the taphonomic problems too complicated.

It would be a mistake, however, to assume that all the complications and depositional behaviour seen at Conisbrough Farms would always occur in every barn, stable and hay loft. In particular, a better drained deep litter bed might develop a distinct foul matter stable fauna. Similarly, midden deposits can develop that contain only large numbers of foul matter species without the "sweet compost" fauna seen here. An example of this is a large midden of horse manure at Sheep Hill Farm, South Yorkshire which contained only *Cercyon* species, predatory staphylinids and a large number of individuals of the *Anthicus* species (Smith 1988).

It must be stressed also that other orders of insects, such as the flies, the lice and the fleas are potentially useful when attempting to identify stable deposits. Kenward and Hall 1997, Hall and Kenward 1998.

Lastly, to dwell on the negative here would be to run the risk of throwing the entomological baby out with the bath water. Despite the fact that it appears beetles may not provide us with the fine levels of resolution that we desired on small farm sites, they are still very informative in terms of issues such as biogeography, the environment around the site, the nature of living and sanitary conditions and the state of decay of materials. From the faunas presented above we would still have been able to suggest that hay was probably present at Conisbrough Park farms and that many of the buildings and surrounding yards contained large amounts of wet, foul and rotten plant materials. Not a bad reconstruction of the actual situation.

ACKNOWLEDGEMENTS

I would like to thank Paul Buckland for realising the potential of Conisbrough Parks Farms for study. Jon Sadler, Mark Dinnin and Pat Wagner all happily dug their way though two metres of cow manure on the odd occasion. I would like to particularly thank Jon Sadler for his many good suggestions on the proof of this paper and Wendy Smith for proof reading. This paper is dedicated to all Dyslexics.

BIBLIOGRAPHY

Buckland P.C., Sveinbjarndóttir G., Savory D., McGovern T.H., Skidmore P., and Andersen C. 1983. Norsemen at Nipaitsoq, Greenland: a Palaeoecological Investigation. *Norwegian Archaeological Review*. 16. 86–97.

Buckland P.C., Sadler J.P. and Smith D.N. 1984. The Beetle's Eye View of the Norse Farm. In: C.D. Morris (ed.). *The Viking Age in Caithness*. Edinburgh University Press.

Coombes C.W. and Freeman J.A. 1956. The Insect Fauna of an Empty Granary. *Bulletin of Entomological Research*. 46. 399–417.

Coope G.R. and Osborne P.J. 1968. Report of the Coleopterous Fauna of the Roman Well at Barnsley Park, Gloucester. *Transactions of the Bristol and Gloucester Archaeo-logical Society*. 86. 84–87.

Green J. 1953. The Beetles of a Cheshire Farm. *Entomologist's Monthly Magazine*. 89. 81–86.

Grieg J.R.A., Girling M.A. and Skidmore P. 1982. The Plant and Insect Remains. In: Barker P. and Higham R. *Hen Domen a timber Castle in the English-Welsh Border.* The Royal Archaeological Institute.

Hall A.R. and Kenward H.K 1990. Environmental Evidence from the Colonia. *The Archaeology of York* 14/6. Council for British Archaeology.

Hall, A.R. and Kenward, H.K. 1998. Disentangling dung: Pathways to stable manure. *Environmental Archaeology.* 1. 123–126.

Hunter F.A., Tulloch B.M. and Lambourne M.G.. 1973 Insects and Mites of Maltings in the East Midlands of England. *Journal of Stored Products Research.* 9. 119–141.

Kenward, H.K. 1975. Pitfalls in the Environmental Interpretation of Insect Death Assemblages. *Journal of Archaeological Science.* 2, 85–95.

Kenward H.K. 1978. The Analysis of Archaeological Insect Assemblages: A New Approach. *The Archaeology of York* 19/1. Council for British Archaeology.

Kenward H.K. 1982. Insect Communities and Death Assemblages, Past and Present. In: Hall A.R. and Kenward H.K. (eds.) *Environmental Archaeology in the Urban Context.* London: Council for British Archaeology. p. 71–78.

Kenward, H.K. and Hall, A.R. 1997. ENhancing Bio-archaeological interpretation using Indicator groups: Stable Manure as a Paradigm. *Journal of Archaeological Science.* 24. 663–673.

Kenward H.K., Hall A.R. and Jones A.K.G. 1980. A Tested Set of Techniques for the extraction of Plant and Animal Macrofossils from Waterlogged Archaeological Deposits. *Scientific Archaeology.* 22. 3–15.

Le Clercq J. 1946. *Insects Brought with hay from the meadows into the hay loft.* Entomologist's Monthly Magazine. 82. 138.

Lucht W.H. 1987. *Die Käfer Mitteleuropas.* Katalog. Goecke & Evers, Krefeld.

McGovern T.H., Buckland P.C., Savory D., Sveinbjarnardóttir G., Andersen C. and Skidmore P. 1983. A study of the faunal and floral remains from two Norse Farms in the Western Settlement Greenland. *Arctic Anthropology.* 20. 93–120.

Perry D.W., Buckland P.C. and Snaesdóttir M. 1985. The Application of Numerical Techniques to Insect Assemblages from the Site of Stóraborg, Iceland. *Journal of Archaeological Science.* 12. 335–345.

Smith D.N. 1991. *An Investigation of the Potential of Modern Analogue Faunas to act as Comparisons to Palaeoentomological samples from Archaeological Farm Sites.* Unpublished Ph.D. thesis. University of Sheffield.

Smith, D.N. 1998. Beyond the barn beetles: difficulties in using some Coleoptera as indicators for stored fodder. *Environmental Archaeology.* 1. 63–70.

Smith I.R. 1989. *Modern Beetles and the Roman Landscape- A Practical Study Using Analogy.* Unpublished BA dissertation. University of Sheffield.

Table 8.1: The Coleoptera from the hay samples at Conisbrough Parks Farm.

	Dec. 1989	Jan. 1989	Feb. 1989	Mar. 1989	Apr. 1989	May 1989	June 1989
Carabidae							
Trechus spp.	-	1	-	-		-	-
Bembidion obtusum Serv.	-	-	-	-	2(1)	-	-
Ptiliidae							
Ptilium spp.		1	-	-	-	-	-
Acrotrichis spp.	-	-	-	-	-	-	1
Staphylinidae							
Micropeplus staphylinoides (Marsh.)	-	2	-	-	1(1)	-	-
Pseudopsis sulcata Newn.	-	1	-	-	-	-	-
Omalium spp.	1	-	-	-	-	-	-
Xylodromus concinnus (Marsh.)	2	4(1)	5(1)	-	1	22(11)	-
Oxytelus spp.	-	-	-	-	1	-	-
Gyrohypnus spp.	-	-	1	-	-	-	-
Xantholinus linearis (Ol.)	-	-	-	-	-	1	-
Philonthus spp.	-	-	1	-	-	1	-
Quedius spp.	-	-	1	-	-	3	-
Tachyporus spp.	1	-	-	-	-	-	-
Aleocharinae Gen. & spp. indet.	-	-	2(1)	-	-	-	-
Nitidulidae							
Glischrochilus quadripunctatus (L.)	-	-	1	-	-	-	-
Cucujidae							
Ahasversus advena (Watl.)	-	-	3(3)	1(1)	2(2)	18(6)	-
Laemophloeus feffugineus (Steph.)	-	-	-	-	-	1	-
Cryptophagidae							
Cryptophagus scanicus (L.)	1	1(1)	3(3)	-	-	2(2)	-
C. spp.	4	6	-	1	-	4	-
Atomaria spp.	-	1	-	-	1	4(2)	-
Lathridiidae							
Lathridius bifasciatus Rtt.	5(2)	3	1(1)	-	-	1	-
Enicmus minutus (L.)	7(2)	1	49(47)	1(1)	20(16)	25(12)	1
E. pseudominutus Strand.	12(2)	-	7(5)	41(22)	-	1	-
E. spp.	9	-	-	-	-	-	-
Cartodere ruficollis (Marsh.)	17(7)	1	168(122)	-	38(30)	824(680)	5(3)
Corticaria spp.	-	-	-	-	4(4)	-	-
Mycetophagidae							
Typhaea stercorea (L.)	9	1	32(6)	-	22(11)	273(180)	12(7)
Anthicidae							
Anthicus floralis (L.)	1(1)	2(1)	1	1(1)	1(1)	7(5)	-
A. formicarius (Goeze)	-	1	1(1)	-	-	-	-
A. spp.	3	3	2	-	1	-	-
Tenebrionidae							
Tenebrio molitor L.	-	-	-	-	1	-	-
Chrysomelidae							
Psylliodes chrysocephala (L.)	-	1	-	-	-	-	-
Curculionidae							
Sitona spp.	-	1	-	-	-	-	-

Table 8.2: The Coleoptera from the top of the straw stacks at Conisbrough Parks Farm.

	Dec. 1989	Jan. 1989	Feb. 1989	Mar. 1989	Apr. 1989	May 1989	Aug. 1989	Oct. 1989	Nov. 1989
Carabidae									
Clivina fossor (L.)	-	-	-	-	-	-	-	-	1
Trechus quadristriatus (Schrk.) or *T. obtusus* Er.	-	2	-	1	-	-	-	-	1
Hydrophilidae									
Cercyon analis (Payk.)	-	-	-	-	1	-	-	-	-
C. spp.	-	1	1	1	-	-	1	1	-
Ptiliidae									
Ptilium spp.	-	4	-	1	-	-	-	-	1(1)
Acrotrichis spp.	1(1)	-	18(4)	8(4)	7(5)	20(4)	11(4)	-	8(6)
Staphylinidae									
Micropeplus staphylinoides (Marsh.)	-	4	4(1)	6	5(1)	5(1)	2	-	8(2)
Xylodromus concinnus (Marsh.)	-	2	2	-	-	-	1	1	2
Oxytelus sculpturatus Grav.	-	-	-	1	1	-	-	-	-
O. sp.	-	-	-	-	-	-	1	-	-
Stilicus orbiculatus Er.	-	-	-	-	-	-	-	1	-
Gyrohypnus punctulatus (Payk.)	-	-	-	-	-	-	-	1	-
Xantholinus spp.	1	-	-	-	-	-	-	1	1
Othius spp.	-	-	-	-	1	-	1	-	-
Philonthus spp.	-	1	-	-	-	1	-	-	-
Quedius spp.	1	-	-	1	-	-	1	1	1
Tachinus spp.	-	-	1	-	-	-	2	-	-
Tachyporus hypnorum (F.)	-	-	-	-	-	-	-	1	-
Tachyphorus chrysomelinus (L.)	-	-	-	-	-	1	-	-	-
Aleocharinae Gen. & spp. indet.	1	3	5	2	-	1	2	-	1
Cucujidae									
Monotoma picipes Hbst.	-	-	-	1(1)	-	-	-	-	-
Cryptophagidae									
Cryptophagus acutangulus Gyll.	-	-	-	1	-	-	-	-	1
C. spp.	-	-	1	-	-	-	-	2	-
Atomaria sp.	-	-	-	-	-	-	-	-	1
Lathridiidae									
Lathridius bifasciatus Rtt.	-	1	-	-	-	1	1	1	1
L. nodifer Westw.	-	-	-	-	-	-	-	1	-
Enicmus sp.	-	-	-	-	-	-	-	1	-
Cartodere ruficollis (Marsh.)	-	-	-	-	-	-	-	1	-
Corticaria spp.	-	-	-	-	-	-	-	6(4)	-
Anthicidae									
Anthicus spp.	-	-	-	-	-	-	-	1	1
Chrysomelidae									
Psylliodes chrysocephala (L.)	-	-	-	-	-	-	-	-	1
Curculionidae									
Ceutorhynchus pollinarius (Frost.)	-	-	-	-	-	-	-	-	1

Table 8.3: The Coleoptera from the deep litter bed at Conisbrough Parks Farm.

	Dec. 1988			Jan. 1989			June 1989		
	DL1	DL2	DL3	DL1	DL2	DL3	DL1	DL2	DL3
Carabidae									
Notiophilus sp.	-	1	-	-	-	-	-	-	-
Trechus quadristriatus (Schrk.) or									
T. obtusus Er.	-	-	1	1	-	1	-	-	-
Pterostichus sp.	-	1	-	-	-	-	-	-	-
Hydrophilidae									
Cercyon impressus (Sturm)	-	1	-	-	-	-	-	-	-
C. sp.	-	-	-	-	-	-	-	-	1
Ptiliidae									
Ptilium spp.	-	-	-	-	-	-	2	-	-
Staphylinidae									
Omalium sp.	-	-	-	-	-	-	-	-	1
Xylodromus concinnus (Marsh.)	-	-	-	1	-	-	-	1	1
Oxytelus sculpturatus Grav.	-	1	-	-	-	-	-	-	-
Gyrohypnus punctulatus (Payk.)	-	-	-	-	-	-	-	1	-
Philonthus laminatus (Creutz.)	-	1	-	-	-	-	-	-	-
P. sp.	-	1	-	-	-	-	-	-	-
Quedius spp.	-	1	1	-	-	1	-	-	-
Tachinus rufipes (Geer)	-	-	1	-	-	1	-	-	-
Nitidulidae									
Meligethes spp.	-	-	1	-	-	1	-	-	-
Cucujidae									
Monotoma sp.	-	1	-	-	-	-	-	-	-
Cryptophagidae									
Cryptophagus spp.	-	2	-	-	-	-	-	-	2
Lathridiidae									
Lathridius bifasciatus Rtt.	-	-	-	-	-	-	-	-	1
Enicmus pseudominutus Strand.	-	6	1	-	-	-	-	-	1
Cartodere ruficollis (Marsh.)	-	-	3	-	-	-	1	2	17
Mycetophagidae									
Typhaea stercorea (L.)	-	1	3	-	-	-	1	2	43
Endomychidae									
Mycetaea hirta (Marsh.)	-	1	1	-	-	-	-	-	-
Anthicidae									
Anthicus spp.	-	1	1	-	-	-	-	-	-
Chrysomelidae									
Psylliodes chrysocephala (L.)	-	1	1	-	-	-	-	-	-
Curculionidae									
Ceutorhynchus sp.	-	-	1	-	-	-	-	-	-

Table 8.4: The Coleoptera from the old midden at Conisbrough Parks Farm.

	Jan. 1989	Feb. 1989	Mar. 1989	May 1989	June 1989	July 1989	Aug. 1989
Carabidae							
Nebria brevicollis (F.)	-	-	1	-	-	-	-
Notiophilus biguttatus (F.)	-	-	-	-	-	-	1
Clivina fossor (L.)	-	-	-	1	-	-	-
Bembidion sp.	-	-	-	-	-	1	-
Pterostichus sp.	-	-	-	1	-	-	-
Amara sp.	1	-	-	-	-	-	-
Hydrophilidae							
Cercyon impressus (Sturm)	-	-	1	4	-	-	2
C. unipunctatus (L.)	-	-	1	6	1	3	1
C. spp.	2	2	1	-	3	2	1
Cryptopleurum minutum (F.)	-	-	-	-	1	-	-
Ptiliidae							
Acrotrichis spp.	-	-	-	-	4	12	-
Staphylinidae							
Micropeplus staphylinoides (Marsh.)	-	-	-	-	1	15(1)	1
Omalium spp.	1	-	-	3	-	3	-
Xylodromus concinnus (Marsh.)	-	-	-	-	3	1	-
Lesteva longelytrata (Goeze)	-	-	1	-	-	-	-
Oxytelus sculptus Grav.	-	-	-	4	2	-	1
O. spp.	-	-	-	2	-	2	-
Stilicus orbiculatus Er.	-	-	-	1	-	-	-
Gyrohypnus sp.	-	-	-	-	1	-	-
Xantholinus spp.	-	-	-	1	1	-	-
Philonthus spp.	-	-	1	9	2	1	1
Quedius spp.	-	-	-	4	2	-	1
Tachyporus obtusus (L.)	-	-	-	2	-	-	-
T.sp.	-	-	-	-	1	-	-
Tachinus rufipes (Geer)	-	-	-	2	-	-	-
Aleocharinae Gen. & spp. indet.	-	-	1	1	-	-	-
Rhizophagidae							
Rhizophagus sp.	-	-	-	1	-	-	-
Cucujidae							
Monotoma picipes Hbst.	-	-	-	-	-	1	-
M. sp.	-	-	-	-	1	-	-
Cryptophagidae							
Cryptophagus sp.	-	-	-	-	1	-	-
Atomaria sp.	-	-	-	3	1	3	1
Lathridiidae							
Lathridius lardarius (Geer)	-	-	1	-	-	-	-
L. bifasciatus Rtt.	-	-	-	-	2	-	2
L. nodifer Westw.	-	-	-	-	1	-	-
Enicmus minutus (L.)	-	-	-	2	4	-	-
E. spp.	-	-	-	-	-	1	1
Cartodere ruficollis (Marsh.)	-	-	1	1	-	-	-
Mycetophagidae							
Typhaea stercorea (L.)	-	-	-	-	2	1	-
Coccinellidae							
Propylea quatuordecimpunctata (L.)	-	-	1	1	-	-	-
Thea vigintiduopunctata (L.)	-	-	-	-	1	-	-
Anthicidae							
Anthicus floralis (L.)	-	-	-	-	1	1	-
A. formicarius (Goeze)	-	-	-	-	-	1	-
Anthicus spp.	-	-	-	2	4	1	1
Tenebrionidae							
Blaps mucronata Latr.	-	-	-	1	-	-	-
Scarabaeidae							
Aphodius sp.	-	-	-	1	-	-	-
Curculionidae							
Sitona sp.	-	-	-	1	-	-	-

Table 8.5: The Coleoptera from the active midden at Conisbrough Parks Farm.

	Jan. 1989	Feb. 1989	Mar. 1989	May 1989	June 1989	July 1989	Aug. 1989
Carabidae							
Notiophilus biguttatus (F.)	-	-	1	-	-	-	1
Trechus sp.	-	-	-	-	-	-	1
Bembidion sp.	-	-	-	-	-	2	-
Hydrophilidae							
Cercyon unipunctatus (L.)	-	-	-	1	-	2	1
C. spp.	-	1	1	3	2	-	3
Histeridae							
Carcinops pumilio (Er.)	3(2)	1	-	-	-	-	-
Ptiliidae							
Ptilium spp.	-	-	-	2	-	-	1
Acrotrichis spp.	-	-	-	-	-	2	-
Staphylinidae							
Micropeplus staphylinoides (Marsh.)	-	-	-	-	1	-	1
Xylodromus concinnus (Marsh.)	-	-	-	2	2	-	2
Oxytelus sculptus Grav.	-	-	-	2	1	-	4
O. sculpturatus Grav.	-	-	3	-	-	-	-
O. spp.	-	-	-	4	1	1	3
Xantholinus spp.	-	-	-	1	-	-	1
Philonthus spp.	1	-	1	12	2	15	3
Quedius spp.	1	-	-	-	-	1	-
Tachinus spp.	-	-	-	2	1	-	-
Leucoparyphus silphodes (L.)	-	-	-	1	-	1	-
Aleocharinae Gen. & spp. indet.	-	-	-	-	2	4	3
Cucujidae							
Monotoma picipes Hbst.	-	-	-	1	-	1	-
M. sp.	-	-	-	-	2	4	-
Ahasversus asvena (Watl.)	-	-	1	-	3	-	1
Laemophoeus ferrugineus (Steph.)	-	-	-	-	-	1	3
Cryptophagidae							
Cryptophagus scanicus (L.)	-	-	-	-	2	2	-
C. spp.	1	-	-	2	1	2	1
Atomaria sp.	-	-	-	-	-	1	3
Lathridiidae							
Lathridius bifasciatus Rtt.	2	-	-	-	-	-	-
L. nodifer Westw.	-	-	-	-	-	-	1
Enicmus minutus (L.)	-	1	2	10	14	8	2
E. pseudominutus Strand.	-	-	-	-	4	-	-
E. spp.	-	-	-	-	-	1	-
Cartodere ruficollis (Marsh.)	2	3	12	198 (2)	397(3)	208(2)	2
Mycetophagidae							
Typhaea stercorea (L.)	1	-	1	15	43	13	-
Endomychidae							
Mycetaea hirta (Marsh.)	-	-	-	1	1	-	-
Coccinellidae							
Propylea quatuordecimpunctata (L.)	-	-	-	-	1	-	-
Thea vigintiduopunctata (L.)	-	-	1	1	-	-	-
Anthicidae							
Anthicus formicarius (Goeze)	-	-	-	-	2	3	14
Anthicus spp.	-	-	-	4	3	5	10
Chrysomelidae							
Psylliodes chrysocephala (L.)	-	-	-	-	-	-	1
Curculionidae							
Sitona spp.	-	-	1	-	-	-	1
Cidnorhinus quadrimaculatus (L.)	-	-	-	1	-	-	-

Table 8.6: The Coleoptera from the yard compost at Conisbrough Parks Farm.

	Dec. 1988	Jan. 1996	Feb. 1996	Mar. 1989	Apr. 1989	May. 1989	June 1989	July 1989	Aug. 1989	Oct.. 1989	Nov. 1989
Carabidae											
Nebria brevicollis (F.)	-	-	1	-	-	-	-	1	2	-	-
Notiophilus biguttatus (F.)	-	-	-	-	1	-	1	1	1	-	1
Trechus spp.	1	1	2	2	2	2	4	2	1	2	1
Bembidion sp.	-	-	1	-	-	-	1	1	-	-	-
Pterostichus sp.	-	-	-	1	-	-	-	-	1	-	-
Agonum spp.	-	-	1	-	-	-	1	-	-	1	-
Amara sp.	-	1	1	1	2	1	1	1	-	1	1
Dromius lineraris (Ol.)	1	-	-	-	-	-	-	-	-	-	-
Hydraenidae											
Helophorus spp.	-	1	1	3	1	3	1	1	1	1	1
Hydrophilidae											
Sphaeridium sp.	-	-	-	-	1	-	-	-	-	-	-
Cercyon impressus (Sturm)	2	-	-	-	-	-	3	-	1	-	-
C. haemorrhoidalis (F.)	2	-	-	-	1	-	-	1	-	-	-
C. unipunctatus (L.)	-	-	-	-	-	-	1	1	-	-	2
C. analis (Payk,)	2	-	-	-	1	-	-	-	-	-	-
C. spp.	14	7	7	20	10	13	22	15	2	6	3
Cryptopleurum minutum (F.)	-	-	-	1	1	1	2	-	2	-	-
Hydrobius fuscipes (L.)	-	-	-	-	-	-	-	-	1	1	-
Histeridae											
Carcinops pumilio (Er.)	-	-	-	-	1	1	1	1	-	1	1
Catopidae											
Choleva spp.	-	-	-	-	-	-	1	1	1	1	-
Orthoperidae											
Corylophus sp.	-	-	-	-	-	-	1	-	-	-	-
Ptiliidae											
Ptilium spp.	9	-	-	-	-	-	-	-	2	1	2
Staphylinidae											
Micropeplus staphylinoides (Marsh.)	-	-	1	1	-	1	3	-	-	-	-
Pseudopis sulcata Newn.	-	1	-	-	-	-	-	-	1	-	-
Proteinus ovalis Steph.	-	-	-	-	-	-	1	-	-	-	-
Omalium excavatum Steph.	-	-	-	-	-	-	-	1	-	-	1
O. spp.	-	-	-	3	2	2	3	3	4	2	3
Xylodromus concinnus (Marsh.)	2	1	1	1	-	2	3	4	3	1	3
Lesteva longelytrata (Goeze)	4	-	2	7	5	5	23	26	6	2	1
Oxytelus sculptus Grav.	-	1	-	2	3	1	14	11	5	5	1
O. rugosus (F.)	1	-	-	2	-	1	1	-	-	-	-
O. sculpturatus Grav.	3	-	-	-	2	-	-	3	-	-	1
O. spp.	-	2	1	5	3	7	13	13	3	-	3
Stenus spp.	1	-	-	1	1	1	2	1	1	-	1
Platystethus arenarius (Fourc.)	-	1	-	-	-	2	2	1	2	1	-
Stilicus orbiculatus Er.	-	-	-	-	-	-	1	-	-	-	-
Lathrobium fulvipenne (Grav.)	-	-	-	-	-	-	1	-	-	1	-
Gyrohypnus punctulatus (Payk.)	5	-	2	2	3	5	5	6	4	2	3
Xantholinusglabratus (Grav.)	-	-	-	-	-	-	-	-	-	-	1
X. spp.	-	1	1	1	1	3	2	-	1	3	2
Philonthus spp.	6	2	3	8	5	5	9	9	9	4	4
Quedius spp.	1	3	3	4	6	7	8	13	4	2	3
Tachyporus obtusus (L.)	1	-	-	-	-	-	-	-	1	-	-
T. chrysomelinus (L.)	-	-	-	-	-	-	1	-	-	-	-
T.sp.	-	-	-	-	-	-	-	-	-	1	-
Tachinus subterraneus (L.)	-	-	-	-	-	-	-	1	-	-	-
T. rufipes (Geer)	-	-	-	-	-	2	-	-	1	-	-
T. spp.	-	1	-	1	1	1	-	2	-	-	-
Leucoparyphus silphodes (L.)	-	-	-	-	-	-	1	1	-	-	1
Aleocharinae Gen. & spp. indet.	5	1	-	5	2	1	4	7	5	3	1
Nitidulidae											
Meligethes spp.	-	-	-	-	-	-	1	1	-	-	-
Rhizophagidae											
Rhizophagus sp.	-	1	-	-	1	-	-	-	1	-	-
Cucujidae											
Monotoma picipes Hbst.	-	-	-	-	-	-	2	1	1	2	1
M. sp.	-	-	-	-	-	-	3	1	3	-	-
Ahasversus advena (Waltl.)	-	-	-	-	-	-	-	-	-	1	-

Table 8.6 continued.

	Dec. 1988	Jan. 1996	Feb. 1996	Mar. 1989	Apr. 1989	May. 1989	June 1989	July 1989	Aug. 1989	Oct.. 1989	Nov. 1989
Cryptophagidae											
Cryptophagus acutangulus (Gyll.)	-	-	-	1	-	-	-	-	-	-	-
distinguendus Sturm.	1	1	-	-	-	1	1	4	1	1	1
C. scanicus (L.)	-	-	1	1	-	1	1	1	1	1	-
C. spp.	6	1	2	7	1	2	8	12	4	2	-
Atomaria spp.	-	-	1	1	1	3	5	6	-	1	-
Lathridiidae											
Lathridius lardarius (Geer)	-	-	-	-	1	1	-	1	-	-	2
L. bifasciatus Rtt.	-	-	-	1	-	-	1	1	-	5	1
L. nodifer Westw.	-	1	-	-	-	-	1	1	-	-	1
Enicmus transversus (Oliv.)	-	-	-	-	-	1	4	1	2	3	2
E. minutus (L.)	2	-	-	1	-	-	2	4	2	-	-
E. pseudominutus (Strand)	-	-	-	1	2	-	48	11	10	3	-
Cartodere ruficollis (Marsh.)	1	3	-	11	3	12	48	11	10	3	-
Corticaria spp.	-	-	-	-	-	-	1	2	-	-	-
Mycetophagidae											
Typhaea stercorea (L.)	1	-	3	5	4	10	20	13	12	21	14
Endomychidae											
Mycetaea hirta (Marsh.)	1	-	-	-	2	-	3	1	-	-	-
Coccinellidae											
Propylea quatuordecimpunctata (L.)	-	-	-	-	-	-	-	2	-	-	-
Thea vigintiduopunctata (L.)	-	-	-	-	-	-	-	1	-	-	-
Anobiidae											
Ptinidae											
Niptus hololeucus (Fald.)	-	-	-	-	-	-	1	-	-	-	-
Tipnus unicolor (Pill. Mitt.)	-	-	-	-	-	-	6	1	1	1	1
Ptinus fur (L.)	-	-	-	-	-	-	-	3	-	-	1
Anthicidae											
Anthicus floralis (L.)	1	-	-	-	-	3	6	-	2	1	-
A. formicarius (Goeze)	-	2	1	1	3	2	3	5	3	4	-
Anthicus spp.	-	3	5	1	3	2	7	7	8	4	4
Scarabaeidae											
Colobopterus fossor (L.)	-	-	-	-	1	-	-	-	-	-	-
Chrysomelidae											
Psylliodes chrysocephala (L.)	-	-	-	-	-	-	1	1	2	2	-
Curculionidae											
Apion sp.	-	-	-	-	-	-	-	-	1	-	-
Phylobius spp.	-	-	-	-	-	-	-	-	1	-	-
Sitona lineatus (L.)	1	-	-	-	-	-	-	-	-	-	-
S. spp.	-	-	-	1	1	1	1	2	1	1	-
Cidnorhinus quadrimaculatus (L.)	1	-	-	-	-	-	1	1	-	-	-
Ceutorhynchus pollinarius (Frost.)	-	-	-	-	-	1	-	-	-	-	-
Sitophilus granarius (L.)	-	-	-	-	-	1	-	-	-	-	-

9. Experimental taphonomy

Louise van Wijngaarden-Bakker

1. INTRODUCTION

Bone weathering is one of the taphonomic processes that skeletal elements undergo in the field. Weathering can be defined as "the natural decomposition of bone and tooth through physical and chemical processes operating at or near ground surface. It involves microscopic changes including cracking, flaking, splitting and fragmentation, and macroscopic changes including breakdown of organic components (chiefly collagen) and dissolution, recrystallization, and/or chemical alteration of mineral components. These changes occur progressively, starting at the time of death, and (can) ultimately result in total destruction of the bone and tooth" (Behrensmeyer 1990).

Actualistic observations on bone weathering have focused so far on the process of subaerial weathering (Beherensmeyer 1978, 1990; Gifford 1980; Lyman and Fox 1989). These studies more or less seem to imply that the process of decomposition stops after burial of the bones. That this is far from true is shown by the study of chemical changes in buried bone from archaeological sites (Gordon and Buikstra 1981; White and Hannus 1983; Locock *et al.* 1992). Actualistic study on the rate of decay of buried bone has been carried out in the experimental earthworks at Overton Down and Wareham (Jewell and Dimbleby 1966; Evans and Limbrey 1974), but the necessary control for taxon, skeletal element and micro-environmental context was not always met.

In this paper the first results of an actualistic study involving the monitoring over eight years of the decay of a sheep burial will be presented. The experiment takes place under well controlled micro-environmental circumstances and it is our aim to continue the observations in the future.

To avoid confusion it seems preferable to restrict the term "bone weathering" to subaerial situations of bone decay. For buried bones the term "diagenetic change" is more appropriate.

2. THE EXPERIMENT

The Dutch Society for Rare Breeds (Stichting Zeldzame Huisdierrassen) has, under its supervision, a number of Breeding Centres that work with recognized breeding standards. One of these centres is "De Breidablik" at Tilburg, where a flock of *ca.* 60 sheep of the "Drentse heideschapen" breed is kept. Throughout the year the flock consists of *ca.* 60 ewes and 10 rams.

Until the introduction of artificial fertilizers "Drentse heideschapen" were kept in large flocks, mainly in the province of Drente. The skeleton of recent specimens strongly resembles that of prehistoric sheep from the Netherlands (Reitsma 1932). The present day breed is small with a shoulder-height of only 50cm. The rams bear large, spiralled horns and the females are hornless or have small straight horns. The live weight varies between 35kg, for females, and 40kg, for males. The breed is characterised by a low quality fleece of *ca.* 1.5–2kg.

One of the rams, born in spring 1980 in the flock of "De Breidablik", was found dead by the owner on July 25 1985.

The ram had been in a good condition all summer and its exact cause of death remains unknown. There is however a possibility that it was struck by lightning. On July 25 the ram was found dead in the meadow. The animal lay on its back with its hindlegs stretched out towards the fence and the front legs bent as if ready to jump. The death of the animal came in a period of high summer temperatures (max. that day 27.2°C) and the carcass was already in an advanced state of decomposition. The flock had been checked two days earlier, so it is assumed that death took place on either July 23 or 24. In view of the circumstances it was decided by the owner to bury the ram on the spot in a small woodland bordering the meadow.

Through personal contact with the Breeding Centre we learnt about the burial and decided to excavate the animal in order to observe the effects of burial in a sandy soil on the surface condition of the individual bones. Although not the primary aim, some observations on the carcass decomposition were also made. At the same time it was decided to start a small monitoring experiment whereby parts of the skeleton would be reburied and re-excavated.

The first excavation:

With the permission of the landowner the first excavation took place on October 12 1988, *i.e.* 38.5 months after the original burial. Although no markers had been placed on the burial spot recollection by one of the people who had helped with the original burial enabled the burial place to be re-located. The original pit of *ca.* 100cm x 90cm was identified. The upper part of the right horn of the buried ram was found at a depth of *ca.* 30cm below ground level. Continuing excavation revealed the complete animal lying on its left side, with the deepest part at *ca.* 65cm below field level. A recent, but disused, rabbit warren had disturbed the filling of the pit, but did not reach the carcass. In the earth that was removed from the pit in order to reach the carcass, some loose rabbit bones were encountered. The complete skeleton was recovered and it was then decided to rebury the left front and hind legs. The pit was partially refilled to a depth of *ca.* 50cm and the bones were placed in anatomical order on the bottom. The pit was then completely refilled and a sketch was made of the location of the burial.

The second excavation:

The property had meanwhile changed from the original proprietor, the Breeding Centre, to the Water Company of the municipality of Tilburg. Although the new proprietor was initially unaware of the monitoring experiment on his land, when asked, full co-operation was given to continue the experiment. The second excavation took place almost five years after the first, on August 13 1993. Most of the reference points on our sketch of 1988 had been removed in the meantime, but with the invaluable help of the former landowner, the reburial spot was again located. The reburied bones were quickly refound and this time the front leg was removed. The hindleg was left *in situ*, entirely untouched. During excavation a small number of rabbit bones were again encountered in the filling of the pit. The pit was then refilled and a new sketch of the location was made. The Water Company promised to place a plastic marker on the spot.

It is our intention to re-excavate the sheep bones in another five years time.

3. THE ENVIRONMENT

The experiment took place on land south-west of Tilburg, (5°6'E, 51°31'N) at an altitude of *ca.* 13.8m above sea level. The burial is located in a small woodland bordering a (former) meadow. The woodland consists of loosely spaced oak trees (*Quercus robur*) with an undergrowth of bushes, mainly alder (*Alnus glutinosa*) and black cherry (*Prunus serotina*). Quite remarkable is the presence of a single cornelian cherry (*Cornus mas*), a tree that does not belong to the local flora. The soil surface was only sparsely covered with seedlings and some grass.

The local soil consists of coversands to a depth of 2m maximum. These coversands, belonging to the Eindhoven formation, are loamy and contain little to no grit. In these coversands a field podzol has developed with an A/B horizon at a depth of *ca.* 36cm. Underneath the coversands lies the Middle Pleistocene Sterksel formation. The Water Company extract water from an aquifer found at a depth of *ca.* 12m. – there was thus no direct danger of contamination of the drinking water by the burial of the carcass.

Since the Water Company acquired large tracts of land in the neighbourhood, a set of measuring points was installed to monitor the chemical properties of the surface water. The water table lies at a depth of 2–3m in summer and of 0–1m in winter. This means that in wet winters the burial may temporarily have been below the local water table level. One of the measuring points, W15, lies in the immediate vicinity of the burial spot. Through courtesy of the Water Company the latest measurements, of July and October 1991, were made available (Table 9.1). W15 is measured bi-annually, so that more recent data will become available in October 1993.

The data indicate that the sheep has been buried in a slightly acid soil, pH 6.4, and with a low content of heavy metals. In the monitoring experiment the effect of local soil conditions on the hydroxyapatite content of the buried bones is of great importance. In this respect the bones have been buried in an active environment with an Agression Factor towards calcium carbonate of >25 (Table 9.1) and a Calcium Activity Coefficient (CAC) of –1.4 (D. Edelman pers. comm.). Both measurements indicate that, eventually, calcium will leach from the bones into the soil.

The Royal Dutch Meteorological Institute (KNMI) at the Bilt has kindly provided the following information on local climatic conditions. Over the past 30 years, *i.e.* from 1961 to 1990 the average annual rainfall amounted to 762 mm/year. In the same period the mean July temperature was 17.0°C and the mean temperature in January 2.3°C.

Data on the acidity of the rain have been provided by the State Department of Public Health and the Environment (RIVM) at Bilthoven (Table 9.2). In the years 1985 to 1991 the pH of the rain fluctuated around 4.7 both at a local and a national level.

4. RESULTS

At the first excavation, three years after the primary burial of the ram, the carcass was found *in situ*, undisturbed by either root action or rabbit activity. The muscles and

Table 9.1: Chemical data (ppm) from measuring point W15 in the direct neighbourhood of the experimental burial. Data provided by the Municipal Water Company of Tilburg.

Minerals	July 1991	Oct. 1991
Cl^-	54	55
NO_2^-	<0·01	0·01
NO_3^-	<0·5	<0·5
SO_4^{--}	180	190
HCO_3^-	81	90
CO_2	80	85
PO_4^{---}	<0·05	0·05
NH_4^+	0·06	0·35
Fe	12·6	26·5
Mn	0·66	0·80
Ca^+	83	87
Mg^+	7·5	7·4
Na^+	25	22
K^+	2·2	2·0
Heavy metals		
zinc	20	20
copper	<10	<10
chromium	<0·5	1·0
nickel	<5	<5
lead	<1	<1
cobalt	<1	1
cadmium	<0·1	<0·1
arsenicum	2	3·5
selenium	<2	<2
mercury	<0·1	<0·1
barium	65	80
aluminium	15	5
tin	<5	<5
Acidity		
pH	6·40	6·40

Table 9.2: Mean annual value of acidity (pH) of rain at local station 231 (Gilze-Rijen) and national mean. Data provided by the Dutch State Department of Public health and the Environment (RIVM).

year	Gilze-Rijen	national mean
1985	4.67	4.68
1986	4.83	4.74
1987	4.69	4.70
1988	4.69	4.69
1989	4.66	4.67
1990	4.80	4.85
1991	4.75	5.01

internal organs had completely decomposed and all cartilage had disappeared. The horn sheaths were present and unaltered. The skin, notably around the front and hind legs and the head, had disappeared, but around the body the woollen coat had formed a kind of felt layer. Within the rib cage some undigested grass was still present.

The bones were of a light brown colour with somewhat darker patches and a shiny, smooth surface. A few of the leg bones were still slightly greasy.

At the second excavation, eight years after the primary burial, the bones were again found *in situ*, just under the root zone. In general the leg bones now showed a light brown to greyish dull surface. Visible alterations of the bone surface were observed. The bones of the hind leg were studied *in situ*, while those from the front leg were removed to allow a detailed study of their surface modification in the laboratory.

The following observations have been made:

Scapula:

The scapula was lying with its costal surface downwards and its lateral surface upwards. The colour of the bone had remained unaltered in comparison with the right scapula that had been excavated five years previously. No surface modification could be discerned on the costal surface. On the lateral surface, both in the supra- and in the infra-spinous fossa, extensive fine-grained pitting was found to be present. The pitting was more dense towards the vertebral border of the scapula and extended somewhat on the anterior part of the spine. The articular end as well as the posterior side of the spine were found to be unaffected.

Humerus:

The humerus was lying with its medial side underneath and the lateral side upwards. Its light brown colour had not changed in comparison with the right humerus. The surface was still smooth and partly shiny. Some very slight and fine-grained pitting was observed on the lateral side of the diaphysis. Both articular ends, as well as the medial side, were unaffected.

Radius:

The radius was reburied with its antero-lateral side upwards and the postero-medial side downwards. On the, still largely shiny, posterior surface of the diaphysis some slight fine-grained pitting was observed. The anterior side of the bone showed a marked change of colour from light brown at the proximal end towards a light brownish grey at the distal end. At the same time the surface changed from smooth and slightly shiny at the proximal end to dull and somewhat rough at the distal end of the diaphysis. Fine-grained pitting was observed on the smoother proximal part of the diaphysis. Both articular surfaces seemed unaffected.

Ulna:

The ulna had been reburied with its medial side downwards and the lateral side upwards, more or less in anatomical connection with the radius. The colour of the bone had not notably changed during the five further years of burial. The medial surface was, however, still somewhat more shiny than the lateral surface which had become dull in texture, especially on the olecranon. Fine-grained pitting was observed on the lateral surface of the olecranon and on both sides of the shaft just below the semilunar articulation. On the distal part the styloid process, fused with the diaphysis, showed no surface modification.

Metacarpal:

The metacarpal was reburied in anatomical connection with the radius and ulna, with its medial side downwards and the lateral side upwards. The entire diaphysis of the bone had markedly changed in colour and become a dull, lighter yellow to grey. More or less intense pitting was observed on the entire diaphysis, whereas flaking of the surface had started on the mid-diaphysis on the lateral side.

The carpals, phalanges and sesamoids were not reburied in the experiment.

The elements of the front leg were weighed and a loss in weight was observed for each of the elements (Table 9.3). The weight loss was found to decrease from the more proximal to the more distal long bones, *i.e.* from the humerus (9.5%) to the metacarpal (2.8%).

The bones from the hind leg:

These bones were not removed from their secondary burial place but were studied *in situ*. Only cursory observations could be made under dark and cramped circumstances. On the pelvis an area of pitting was observed around the acetabulum, the femur and tibia seemed unaffected, while the surface of the metatarsal exhibited a greyish colour and some superficial flaking, similar to that observed on the metacarpal.

Table 9.3: Bone weight (g) of right and left elements after respectively three and eight years of burial.

Element	right	left	difference	%
scapula	59·8	57·3	2·5	4·2%
humerus	80·8	73·1	7·7	9·5%
radius	56·2	52·9	3·3	5·9%
ulna	16·5	15·6	0·9	5·5%
metacarpal	35·5	34·5	1·0	2·8%

5. DISCUSSION

Bone decay can be described as a series of overlapping reactions that are controlled by water, acid, oxygen and calcium contents in the bone and soil. At first bone collagen is decomposed by microorganisms, whereby hydrogen-enriched hydroxyapatite is formed. These reactions occur inside the bone and are independent of pH. In acid soils the hydrogen-enriched hydroxyapatite will be decomposed into Ca and HPO_4 ions (White and Hannus 1983). However, within a group of securely dated post-medieval bone samples, the relationship between chemical composition of skeletal element, length of burial and soil conditions was found to be strongly non-linear (Locock *et al.* 1992: 300).

The same authors subjectively grouped their bone samples into five classes:

I. Fresh, greasy surface
II. Surface dulled, no longer greasy
III. Some surface deterioration, pitting, powdering
IV. Severe surface deterioration: whole thickness of bone affected, cracking and splitting
V. Disintegration, losing cohesion when handled (Locock *et al.* 1992: 299).

The experimental data assembled in our study give evidence of a somewhat different sequence. The first symptom of diagenetic change was found to be a fine-grained pitting of the otherwise smooth bone surface. The pitting was observed on the flat surface of the scapula and on the diaphyses of the humerus, radius and ulna. On the radius the next stage could be observed, namely the intensification of pitting, which changes the texture of the bone surface to dull and more or less roughened. The next stage of surface modification was seen in particular on the metacarpal where flaking was observed on the mid-diaphysis.

These observations lead to the following sequence of surface modification by diagenetic changes:

A. Fresh, smooth surface
B. Smooth surface with fine-grained pitting
C. Extension of pitting leading to a dull, rough surface
D. Beginning of flaking

Further stages in the surface decomposition may become available at the continuation of our experiment.

A marked decrease in weight of the individual bones has been observed, ranging between 2.8% for the metacarpal and 9.55% for the humerus.

Our experimental observations have revealed a differential rate of surface alteration according to skeletal element. The more distal elements, in particular the metacarpal and metatarsal, were found to be more affected by diagenetic changes than the more proximal elements such as the humerus and femur. The radius and tibia seem to take an intermediate position. The rate of loss in bone weight, however, seems to follow the reverse sequence with the humerus more affected than the metacarpal. One possible explanation for these differential rates may perhaps be found in the different rate of maturation of the skeletal elements. Within the appendicular skeleton there is a gradient of early to late maturation from the feet bones towards the pelvis and scapula (Bergström 1974:14). Data on a possible relationship between the degree of maturation and the proportion of organic versus inorganic components within the different types of bone have not been found in the literature.

In the Tilburg experiment the observed diagenetic changes were more prominently present on the upward oriented surfaces of the buried bones than on their downward oriented surfaces. It is tempting to relate this phenomenon to the influence of acid rain percolating downwards through the soil. As noted above, the local pH of the rain is far more acid (pH 4.7) than that of the soil in which the ram was buried (pH 6.4). More research is needed to establish the proposed relationship between diagenetic change of bone surface and the influence of acid rain. The preferential degradation of the upper surfaces has also been noted in experiments of subaerial bone weathering (Behrensmeyer 1978: 153). At the experimental earthwork at Wareham "in the turf, the bones were found to have lost most of the bone table [*surface layer*: *eds.*] comprising the face that was uppermost. This upper side of the bone had in many places completely dissolved away, exposing the whole of the marrow cavity...In some places solution of the lower side of the bone had begun" (Evans and Limbrey 1974: 193). The pH of the local subsoil on which the earthwork was built at Wareham varied between 4.5 and 5.1, which makes the microenvironment far more acid than that in the Tilburg experiment. Data on the acidity of the rain in the years that the bones were experimentally buried at Wareham (1963 to 1972) could throw some light on this matter.

6. CONCLUSIONS

The monitoring of the diagenetic changes in the separate skeletal elements of a sheep carcass buried under known microenvironmental circumstances has shown a differential rate of surface deterioration of the individual bones.

The bone surface decomposition, as a result of diagenetic changes, follows a general pattern with the following sequence:

A. Fresh, smooth surface
B. Smooth surface with fine-grained pitting
C. Extension of pitting leading to a dull, rough surface
D. Beginning of flaking

Secondly, a relationship was observed between the position of the element within the skeleton and the degree of surface alteration, with the more distal elements more affected than the proximal elements. A tentative relationship with the degree of maturation of the different skeletal elements is proposed.

Finally, the influence of diagenetic change was found more prominently on the upward oriented surfaces of the buried bones than on their downward oriented surfaces.

For each individual long bone the part facing upwards during burial was found to be more affected by diagenetic changes than the downward facing surface. Influence on the diaphyses was also found to be greater than on the articular ends of the long bones. Continuous monitoring and further chemical analysis will be carried out in order to follow the observed processes.

ACKNOWLEDGEMENTS

Most sincere thanks are due to Mrs W.A. van Nunen-Forger, the owner of "De Breidablik" for her invaluable and continuing help in the experiment. Thanks are also due to the Water Company of Tilburg for their permission to carry on the experiment and to Dick Edelman for making available the chemical data from the terrain. Willem Schnitger, Eli Gehasse, Machteld van Dierendonck and Rik Maliepaard assisted in the two experimental excavations.

BIBLIOGRAPHY

Behrensmeyer, A. K. (1978). Taphonomic and ecological information from bone weathering. *Paleobiology* 4, 150–162.

Behrensmeyer, A.K. (1990). Experimental taphonomy workshop. Papers presented at the ICAZ Conference Washington, 1990.

Bergström, P. L. (1974). Groeiritme en karkassamenstelling I. *IVO Rapport* B-118, Zeist.

Evans, J.G. and Limbrey, S. (1974). The experimental earthwork on Morden Bog, Wareham, Dorset, England: 1963 to 1972. *Proceedings of the Prehistoric Society* 32, 170–202.

Gifford, G.P. (1980). Ethnoarcheological contributions to the taphonomy of human sites. pp. 94–107 In: Fossils in the making (eds. A. K. Behrensmeyer and A. K. Hill),. University of Chicago Press, Chicago.

Gordon, C. E. and Buikstra, J. E. (1981). Soil pH, bone preservation, and sampling bias at mortuary sites. *American Antiquity* 46, 566–571.

Jewell, P. A. and Dimbleby, G. W. (1966). The experimental earthwork at Overton Down, Wiltshire, England. *Proceedings of the Prehistoric Society* 32, 313–342.

Locock, M., Currie, C. K. and Gray, S. (1992). Chemical changes in buried animal bone: data from a postmedieval assemblage. *International Journal of Osteoarchaeology* 2, 297–304.

Lyman, R. L. and Fox, G. L. (1989). A critical evaluation of bone weatering as an indication of bone assemblage formation. *Journal of Archaeological Science* 16, 293–317.

Reitsma, G. G. (1932). *Het Schaap. Zoölogisch onderzoek der Nederlandse terpen 1.* Wageningen.

White, E. M. and Hannus, L. A. (1983). Chemical weathering of bone in archaeological soils. *American Antiquity* 48, 316–322.

10. Why did the chicken dig a hole? Some observations on the excavation of dust baths by domestic fowl and their implications for archaeology

Keith Dobney, Allan Hall and Michael Hill

A group of Old English game birds (*Gallus* f. domestic) were introduced to an enclosure constructed of posts and chicken wire in the lee of a tall (3.5 m), North-facing, brick wall in the Walled Garden of Heslington Hall, York, in May 1992. The accompanying photographs and drawings record some of the results of their activities.

Within a few weeks of arrival the eight birds had removed much of the existing vegetation – a mixture of weeds and ornamentals. The area was, at the time of writing, almost barren below the height of an adult cockerel: only a few plants, such as stinging nettle (*Urtica dioica* L.), angelica (*Angelica archangelica* L.), ivy (*Hedera helix* L.) and gooseberry (*Ribes uva-crispa* L.), together with some saplings of hazel (*Corylus avellana* L.) and willow (*Salix* sp.), which were already quite tall when the chickens arrived, have survived.

Over the months, the chickens were observed to create several dust baths (for the purpose of removing ecto-parasites) and, in doing so, excavated hollows of considerable size at the foot of the wall and also at the bases of some of the posts supporting the netting on the top of the run. These hollows became sinks for a variety of debris including some sizeable twigs and stones, as well as bones, feathers, plastic pens, plastic plant labels, and fragments of glass.

During a period in which the population of the run rose to thirteen chicks and five adult birds, one particular depression formed in a matter of only a few weeks. It had a distinctive square-edged wall (Figures 10.1 and 10.2).

Features whose interpretation is unclear are often observed on archaeological sites. There were, for example, a series of shallow 'depressions', not associated with any structures or larger negative features, dated to the early years of Anglo-Scandinavian occupation at 16–22 Coppergate, York (Kenward and Hall, 1995). Thus from the levels dated by R. A. Hall (pers. comm.) to the mid 9th-or late

Figure 10.1: The chicken depression – note accumulation of debris, as of summer 1993.

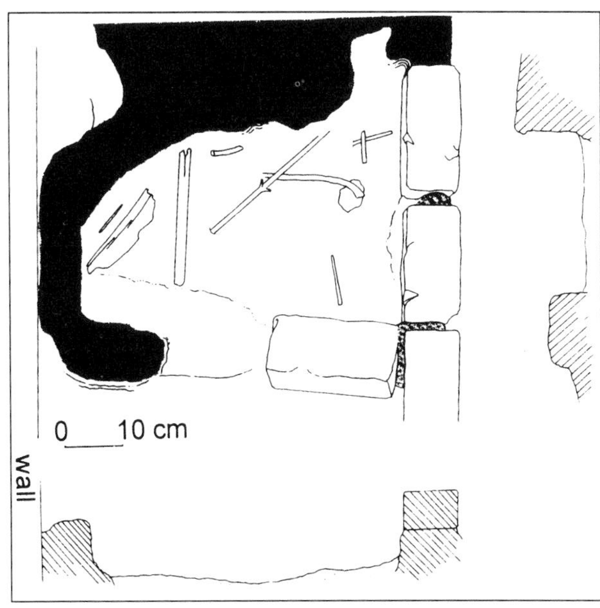

Figure 10.2: Plan of the chicken run, summer 1993.

Figure 10.3 and 10.4: Chicken run in winter 1995 – note extension of depression and 'movement' of brick.

Figure 10.5: Area of feeding within the run – note accumulation of wet, organic material.

9th/early 10th centuries a series of ten cuts defined as 'depressions' was identified. These ranged in lateral extent from about 30 cm to more than 2.5 m, with depths of about 10–20 cm (rarely 30 cm). Some of these features are now thought by the excavator to be the bases of pits truncated by later activity but not recognised as such during excavation. On the basis of our observations, the smaller features may easily have been dust-baths of the kind discussed here, whilst the larger examples may have resulted from the 'wallowing' activities of pigs.

The distinctly 'artificial' appearance of the feature illustrated in Figures 10.1 and 10.2, associated with what are clearly man-made structures (in this case a brick wall, a line of loose bricks originally forming a border to a flower bed, and wooden posts), might well have led to a spurious interpretation if excavated on an archaeological site.

After these features were recorded in the summer of 1993, they were substantially modified by continued use and the effects of precipitation (which to some extent helped to consolidate these rather ephemeral features by compaction of the sediment at the edges). Comparison of Figures 10.1 and 10.2 (summer 1993) with Figures 10.3 and 10.4 (winter 1995) shows how far the depression at the foot of the wall was extended (beyond the single brick at right angles to the line of loose bricks which provided a fixed marker). It will also be noticed how much the angle of rest of the isolated brick changed, and how far the clearly defined 'border' to the left and rear of the depression was remodelled. After the photograph in Figure 10.3 was taken, a prolonged spell of dry weather in March 1995 permitted further dust bathing, resulting in additional changes to the features.

At the opposite end of the run from the area used for dust bathing, other, less well defined features were observed (Figure 10.5). It is in this area that the chickens were routinely fed and the disturbances to the surfaces here were the result of scratching by the fowl during feeding. The area was also regularly waterlogged (perhaps a function of compaction by humans and chickens!) and straw was laid to mitigate this. The matted organic matter and soil rich in straw, uneaten grain and chaff, and hen droppings took on the appearance of what, under other circumstances, might be described as 'stable manure'.

These observations serve to remind us of some of the dangers inherent in interpreting archaeological features and biological remains preserved in archaeological deposits.

BIBLIOGRAPHY

Kenward, H.K. and Hall, A.R. (1995) Biological Evidence from 16–22 Coppergate. The Archaeology of York: The Environment 14/7. York: Council for British Archaeology.

11. Arthropod remains as indicators for taphonomic processes: an assemblage from 19th century burials, Broerenkerk, Zwolle, the Netherlands

Tom Hakbijl

1. INTRODUCTION

Arthropod remains can be very useful tools in the reconstruction of formation- and taphonomic processes. The potential of this approach has not been fully utilised.

An assemblage is presented from a group of 19th century burials in a church: Broerenkerk, Zwolle, the Netherlands. This assemblage can be used as a reference in future research.

2. CONTEXT

In 1987/88 the tombstone floor of the church was excavated. In the sandy subsoil many coffined corpses had been interred, mostly in three layers. The coffins were made of wood, of which a small proportion was partly preserved. Eight of the burials in our study could be identified in a burial register of the period between 1819 and 1828. The two other burials dated from the same period, or earlier.

During the excavation insect remains were found. Larger specimens and concentrations of smaller specimens were sampled.

3. RESULTS

In samples of various sizes, with a total weight of only 90g, and taken from ten burials, remains of nine species of arthropods were found (Table 11.1). *Rhizophagus parallelocollis*, *Conicera tibialis* and the millepedes were found in very high numbers. The other species were less abundant. The recovered species have in common that they are obligatorily or facultatively subterranean. *Conicera tibialis* and *Rhizophagus parallelocollis* are reported to occur regularly in 19th century burials. The habitats of the recovered species are presented in Table 11.2.

Table 11.1: Arthropod remains from burials, Broerenkerk, Zwolle.

ARACHNIDA, PSEUDOSCORPIONIDA
Chernetidae
Pselaphochernes scorpioides (Hermann)
INSECTA, COLEOPTERA
Pselaphidae
Trichonyx sulcicollis (Reichenbach)
Elateridae (Click Beetles)
Melanotus erythropus (Gmelin), or *castanipes* (Paykull)
Rhizophagidae
Rhizophagus parallelocollis (Gyllenhal)
INSECTA, DIPTERA: CYCLORRHAPHA
Phoridae (Humpbacked Flies)
Conicera tibialis (Schmitz) (mainly puparia)
DIPLOPODA (Millepedes)
Blaniulidae
Cylindroiulus britannicus (Verhoeff), or *latestriatus* (Curtis)

add: some unidentified Oribatida (Acari) and Sciaridae (Diptera, Nematocera)

Conicera tibialis (the coffin-fly)

This small fly, which was not described until 1925, is the most common and abundant species on corpses that have been buried for more than a year. Also in the samples from the Broerenkerk it is the most abundant species. *Conicera tibialis* is a specialised feeder on buried corpses

Table 11.2: Ecology of the recovered species

species	food source	substrate
Pselaphochernes scorpioides	other invertebrates	wood, corpses etc.
Trichonyx sulcicollis	mould	wood etc.
Melanotus erythropus/castanipes	detritus from wood	wood
Rhizophagus parallelocollis	mould?	corpses, wood etc.
Conicera tibialis	animal matter	corpses
Nopoiulus kochii	detritus	wood etc.
Cylindroiulus britannicus/latestriatus	detritus	wood etc.

and carrion. It appears to be able to locate a buried corpse and find access to it from above ground. Once the corpse has been colonised, this species can multiply for generations without emerging to the surface. Where burials are concentrated this species can even colonise new corpses without coming to the surface.

Rhizophagus parallelocollis

This species is found in a later stage of decomposition. It is also widespread, but less abundant than *Conicera tibialis*. It can be found on corpses that have been buried for 2 years or longer. Also this species gains access to buried corpses from above ground. They appear to feed on fungi present on older corpses rather than on the corpse itself. They can also be found in cellars on mouldy wood and on plant remains.

Of the other species only *Trichonyx sulcicollis* was previously reported from 19th century burials, but with a much lower frequency. The other arthropods are recorded for the first time, although other species of millepedes than we found in this study were reported earlier. The decaying wood of the coffins seems to be an important resource for these creatures.

4. CONCLUSION AND DISCUSSION

The interment of coffined corpses inside a church gives relatively little opportunity for arthropods to gain access to the corpse. Nevertheless, the number of species appears to be substantial.

In modern and ancient societies very different ceremonies do, or did, exist, for instance, prolonged periods of exposure of the deceased. Also the remains of people who were not buried with the usual care for that particular period may be encountered in archaeology. In such instances many more species may have gained access. Those insects and other arthropods may belong to other ecological groups, for instance:

- moderately well burrowing necrophilous species and
- non-burrowing species specialised on different stages:
  - fresh corpses
  - bloated stage
  - in decay
  - dry stage
- species feeding on mummified corpses
- water insects
- ectoparasites
- stored products species
- displaced specimens.

The same techniques can, of course, also be used to reconstruct taphonomic processes involving animal remains other than human. Even buried organs that leave no remains can, in theory, be traced by using insects as tools.

BIBLIOGRAPHY

Smith, K.G.V. (1986). *A manual of Forensic Entomology*. London: The Trustees of the British Museum (Nat. Hist.).

12. Context level interpretation of animal bones through statistical analysis

Marta Moreno Garcia and James Rackham

1. INTRODUCTION

Statistical analyses have been used by zoo-archaeologists as an aid to the interpretation of animal bone assemblages. In order to be able to apply a statistical technique one needs to quantify the material to be analysed in a suitable way. Some quantification techniques may not fulfill the assumptions required for a particular statistical method.

A variety of quantification methods (Grayson 1984; Ringrose 1993) are in use with one main objective; that is to assess in archaeological terms intra- and/or inter-site variation among animal bone assemblages. To achieve this aim, the faunal analyst must be aware of the nature of the data he or she is intending to quantify, since there are a number of factors such as different recording systems, recovery methods and taphonomic processes that bias the data collected complicating any attempt at valid interpretation both at a statistical and an archaeological level.

The aim of this paper is to show how, using a particular bone recording method (Rackham 1986), it has been possible to quantify vertebrate faunal data that could then be analysed by the Pie-slice program of Orton and Tyers (1990, 1992) originally developed for the statistical analysis of ceramic assemblages.

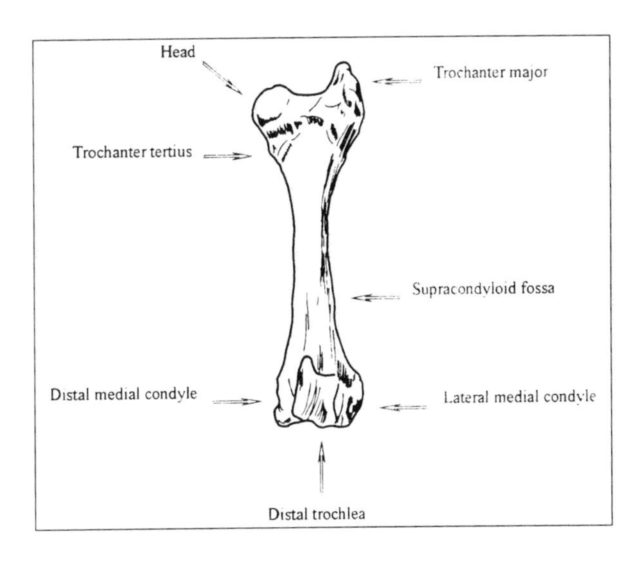

Figure 12.1: 'Diagnostic zones' – the femoral zones

2. THE DATA

The data were taken from the archives of the (then) Greater London Environmental Archaeology Service of the Museum of London, where they had been recorded using the method Rackham (1986) developed based on Watson's (1979) concept of 'diagnostic zones'. According to this method each skeletal element has a number of identifiable zones (for an example, see Figure 12.1) that can be quantified to consider fragmentation and bone representation within individual contexts or groups. The Pie-slice program (Orton and Tyers 1990) seemed applicable to animal bone data recorded under this quantitative method.

A preliminary study was carried out by Moreno Garcia (1991) on Roman, Saxon, Medieval and post-medieval bone assemblages (mainly from London). Contexts with more than 200 bones were selected from the data recorded for a number of sites. It was considered that this was an appropriate number to ensure that samples were not too small for archaeological interpretation and/or statistical analysis.

This paper only deals with the Roman contexts from the ten sites listed in Appendix 12.1. The first field of the dataset consists of the site code and the chosen contexts. The second and third fields are species and skeletal element respectively. The species and element codes used are listed in appendices 12.2 and 12.3, respectively. The species include cattle, sheep/goat, sheep, pig and horse, as well as cattle-size and sheep-size where the precise species could not be identified. Most skeletal elements are included except for ribs, the patella and small elements from the feet and ankles.

3. THE ANALYSIS

Using all the 'diagnostic zones' of the same skeletal element in a single context a measure of the proportion of each bone present in that context can be obtained (Rackham 1986). This measure is transformed by the Pie-slice program into counts (for more details of the program see Orton (forthcoming) and Moreno Garcia *et al.* (forthcoming).

The transformed data are then structured in a three-way table, in which the variables are: site/context, species and skeletal element. Fragments with no zones do not appear in the data set, and an assumption is made that if fragments of the same bone element and species do not join together physically they derive from different bones.

To analyse the data, two statistical techniques are employed: quasi-log-linear analysis (Bishop *et al.* 1975) and correspondence analysis (Greenacre 1984).
The former detects any interactions between the three variables. An interaction occurs when the value of one variable is associated with the value of another. The results are presented as tables.

The latter technique is used to display graphically those interactions shown in the quasi-log-linear analysis tables. Correspondence analysis (CA) displays the values of two variables in the same graph. For instance, the "species by bone" analysis plots the distribution of the proportions of skeletal elements in each of the species. Thus, those skeletal elements associated with a certain species are plotted close to it. In addition, those species with equal or similar proportions of bone are said to have the same profile and are merged into the same cluster. This is indicated by the tilda symbol (~) in front of one of the species merged. The rest go in brackets and do not appear on the plot.

In summary, correspondence analysis (CA) provides a way of re-expressing the original data in a more comprehensible form for archaeological interpretation. That is the reason why only the graphics obtained after carrying out CA on our Roman dataset are presented and discussed in the next section.

4. RESULTS OF THE ANALYSES AND THEIR DISCUSSION

The first CA displays the interactions between **species** and **bone**, that is to say how different skeletal elements are differentially represented for different species, irrespective of context (Figure 12.2).

The first axis (horizontal) shows the contrast between

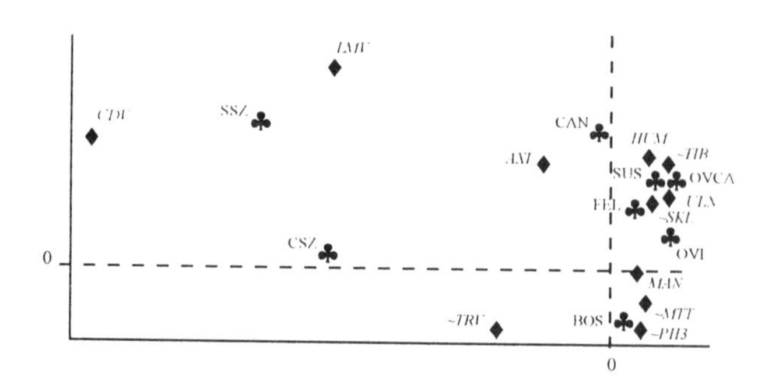

Figure 12.2: CA plot for species (♣) and bone element (♦).

cattle-size (CSZ) and sheep-size (SSZ) and the other species, and between the four types of vertebrae (CDV, LMV, ~TRV (CEV)) and the other bone elements. In other words, there is a high proportion of vertebrae associated with the 'size' categories that could not be identified to species, and generally an under-representation of these skeletal elements for the other species. It seems we are getting a pattern that could be due to a problem of archaeological identification or recording. Vertebrae from the dataset were not generally identified to species level but in relation to their size were recorded under the general categories, cattle-size (CSZ) and sheep-size (SSZ).

The view on the second axis could give us additional information but due to the strong interaction present on the first axis this is quite difficult to see. However, there is a contrast between cattle (BOS) and ovicaprid (OVCA) plus pig (SUS), and between the metatarsal group [~MTT including AST, ATL, CAL, PH1, SCP and MTC)], the phalange group [PH3 and PH2], plus the thoracic/cervical vertebrae [TRV and CEV] on the one hand and the humerus [HUM] and tibia [TIB with FEM and RAD] on the other. In this case, most of the variation has been caused by the way part of the data has been recorded. To improve the picture and clarify the pattern showing up in the vertical axis either a) CSZ, SSZ and the vertebrae can be removed from the data or, b) CSZ can be merged with BOS and SSZ with OVCA (and OVI), and the analyses re-run. The latter was preferred.

The new plot (Figure 12.3) now shows two distinct clusters on the first axis. One with cattle (BOS) and a thoracic vertebrae group [TRV with CEV and PH3] plus a phalange group [PH2 with AST, ATL, PH1, MTC and MTT) and another with sheep [OVI], sheep/goat [OVCA] and pig [SUS] with the radius, humerus, tibia and ulna [RAD, HUM, TIB and ULN].

The plot shows that vertebrae and phalanges are over-represented for cattle whilst limb bones are lacking for this species. In contrast, the limb bones are strongly associated with ovicaprids and pig.

The pattern emerging confirms the view of the second axis in the previous run. These results led us to look for different archaeological interpretations. On the one hand, the fact that "small bones" and fragments (phalanges and

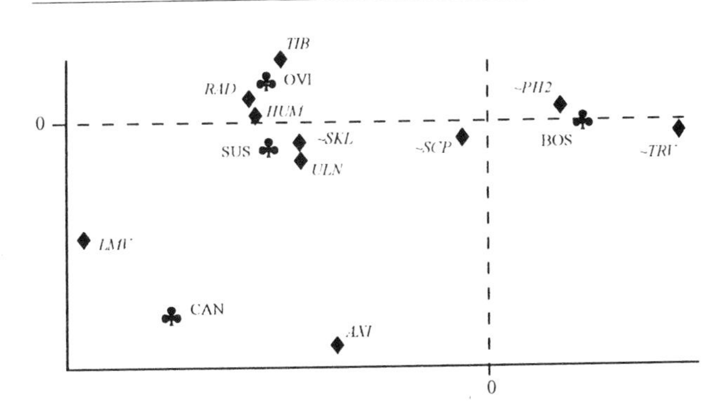

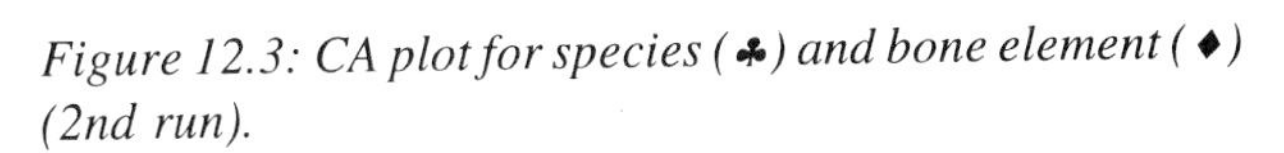
Figure 12.3: CA plot for species (♣) and bone element (♦) (2nd run).

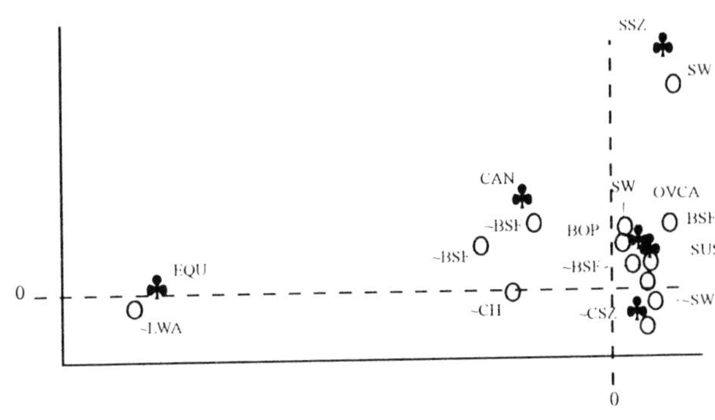

Figure 12.4: CA plot for context (O) and species (♣).

vertebrae) are more numerous for the bigger of the species (cattle) could be pointing to poor recovery. Small bones of large species tend to be recovered more often than the smaller bones of small species which are easily left in the ground when appropriate excavation techniques (i.e. dry and wet sieving) are not used (Payne 1972). However, although this could account for the absence of phalanges and vertebrae of ovicaprids in some of our sites it does not account for the lack of cattle long bones.

On the other hand, natural taphonomic processes cannot be forgotten (Gifford 1981). Under the same preservation conditions cattle limb bones are meant to be better preserved than ovicaprids bones. However, it is also true that bones of the smaller species tend to fragment into fewer pieces than the bones of much larger species. If this had happened, cattle long bones may have been too fragmented to be zoned and consequently would not be included in the analysis. It was necessary to check the original data to see what differences there were, if any, between sheep/goat long bone fragments [OVCA LBF] and cattle long bone fragments [BOS LBF]. The data showed that there were more BOS LBF than OVCA LBF. Thus, the lack of limb bones for BOS might be accounted for, in part, by the fragmentation they underwent.

Nevertheless, the cause of the fragmentation does not necessarily have to have been due to 'natural' taphonomic processes. We could be dealing with the results of deliberate butchery, particularly for marrow (Armitage 1982) or even grease extraction (Maltby 1989, 90). Due to the smaller size of ovicaprid and pig bones, their carcasses may not have received a treatment similar to the larger limb bones of cattle (Maltby 1984). Thus the bones of SUS, OVI and OVCA would have been recovered as single, intact or semi-intact elements, whereas it is likely that a high proportion of BOS limb bones consisted almost entirely of small fragments with no zones. This hypothesis could be supported by further data including the physical evidence of butchery, but because this was a preliminary study such information was not included in the analysis. The results are sufficiently encouraging to consider other variables in the analysis that might help with the interpretation. The results of the context by species analysis provided further details that defined more clearly the pattern obtained with this species by bone analysis.

The second analysis was **context** by **species** (Figure 12.4). It is designed to recognise any association between species and context. The first principal axis is dominated by horse and five contexts from three of the London sites [LWA84:373, CH75:5089 and HOO88:454, 590, 634]. In the second axis (vertical), again there is a contrast between cattle (~CSZ) and sheep/goat (OVCA) that cannot be seen properly because horse and the contexts associated with it show such a strong interaction. It is worth mentioning that CA is sensitive to outliers; in other words, the impact from a small number of variables strongly associated, as is here the case between horse and a group of contexts, overshadows the pattern emerging from the remainder of the data. Once these outliers have been noted, it is best to remove them from the data and re-run the analysis. Before discussing the results of the second run it seems appropriate to point out that LWA84:373, CH75:5089 and HOO88:454, 590, 634 correspond to areas of marginal land (Schofield 1987) where dead horses could well have been disposed of. HOO88 is a Roman cemetery.

On the second run with the outliers removed (Figure 12.5), the first axis is dominated by a contrast between ~SSZ (OVCA, OVI) and SUS, and ~CSZ (BOS). Equally there are clusters of contexts associated with the different species. Certain contexts from Stanwick [SW84II:1035, 1027] and Beddington Sewage Farm [BSF86:77803, 77804] and from Bishopgate, London [BOP82] are associated with ovicaprids whilst several contexts associated with various of the London sites, particularly Crutched Friars [RAG82:1549, (HOO88:454, 590, LOW88:1241, 484, LWA84:373, OPT81:112, RAG82:1526), together with one from Stanwick [SW85:2106] are asociated with cattle. On the second (vertical) axis pig stands out, as do context groups BSF86:88602 and ~SW85:2200 from Beddington Sewage Farm and Stanwick.

Before interpreting these patterns two factors should be noted. Firstly, those contexts associated with sheep sized are mainly from rural sites. BSF86 is Beddington

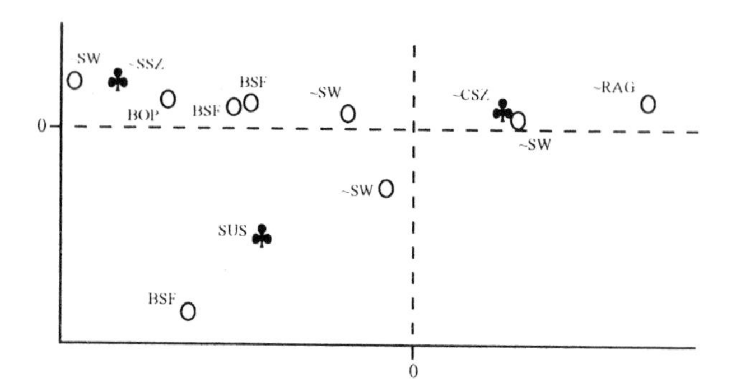

Figure 12.5: CA plot for context (o) and species (♣) 2nd run).

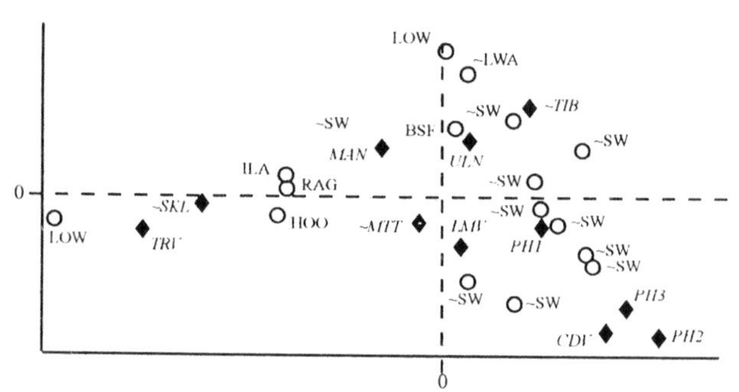

Figure 12.6: CA plot for context (o) and bone element (♦).

Sewage Farm, a villa site in Croydon and SW84II is an Iron Age and Romano British settlement at Stanwick in County Durham, while those associated with cattle sized are from the Roman City in London. Secondly, the BSF86 contexts associated with ~SSZ are dated to late Iron Age and those from Stanwick associated with ~CSZ are dated from the end of the 2nd century onwards.

The way the data have grouped hints at differences between rural and urban economies in Roman Britain (King 1978). Cattle were by far the most important producers of meat. It is therefore to be expected that in the towns their carcasses would have been intensively exploited to maximise their food value and the positive association of ~CSZ (BOS) with the urban sites illustrates the importance of cattle to the urban economy. The results of the analysis show an association between ~SSZ (OVCA, OVI) and the rural sites suggesting that more rural communities may have continued with a pattern of animal husbandry, emphasising sheep, seen in the Iron Age and native sites (King 1978) despite the process of Romanization. This hypothesis must be taken in conjunction with the next analysis "context by bone" to better understand the variability present in the different contexts.

The third analysis was **context** by **bone** (Figure 12.6). The kind of interaction expected in this analysis might show varying proportions of different bone elements in different contexts, irrespective of the species to which they belong.

In the first axis, contexts from Crutched Friars, Cross Key Court and 85 London Walls [~RAG82:1549 (RAG82:1189, 1526, 1533), ~RAG82:1339 (OPT81:112, RAG82:1514, 1079) and LOW88:484] form a cluster associated with a high proportion of cervical, thoracic and skull fragments (TRV, ~SKL). They also show a lack of limb bones. Was it possible that, once more, the absence of the variable LBF (long bone fragment) from the dataset could have biased the result? In order to test this, the number of LBF in the different contexts was checked. The outcome showed that there were no real differences among them and in some cases, RAG82, LOW88 and OPT81 contexts include a lower number of LBF than BSF86, LWA84, BOP82 and ILA79 contexts which are associated with a high proportion of limb bones on the second axis. This result left us to considered another hypothesis.

We could look for a functional interpretation. It is possible that the different proportions of bone elements present in the various contexts reflect intra- and/or inter-site function (Serjeantson 1989). Limb bones carry the best cuts of meat. Thus, once disjointed by the butcher, they might have been distributed among the households of the city leaving behind in the contexts at RAG82, LOW88 and OPT81 bones that may be considered as the butchery waste; that is to say, those parts of the carcass that were discarded at the primary butchery site (Maltby 1989). Some of these contexts are located just within the Roman City Wall, in the so-called "non residential areas", where "industrial" activities such as butchery might have taken place.

The abundance of bones reflecting good meat cuts (such as femur, humerus, tibia and radius) together with a low proportion of primary butchery waste (phalanges, metapodials, vertebrae and skulls) in ~LWA84:373 (BOP82:287, BSF86:77806, ILA79:394) and ~SW85:2200 (BSF86:48303, 59901, 77804) suggests that these contexts are related to the domestic debris derived from meat consumption. Their position in the plot shows the contrast with ~SW85:2106 (CH75:3089, BSF86:88601, SW84II:1001, SW85:2012) which are heavily associated with all the phalanges. It is possible that the presence of phalanges and the lack of any other bones in these contexts point, not to butchery activities, but to tanning (Serjeantson 1989, 5).

Although these results seem to point to real functional variations among the various contexts, one cannot forget that some of this variation can be accounted for by differential preservation, retrieval strategies and even level of identification.

5. CONCLUSION

Taking together the results of the three analyses, one can see that the Pie-slice program has proved very useful in pointing out the differences not only among sites but also between contexts. It seems possible to distinguish contexts related to particular activities in terms of the proportions of the different skeletal elements and species present. At the same time, the interactions between all the variables make us aware of biases that can be included in a dataset, particularly in the early stages of recovery and recording the data.

Further research is needed and other variables such as age, sex and type of context ought to be included in the analysis to confirm some of the hypotheses mentioned above. None the less, this preliminary study shows the great potential that the application of the Pie-slice program has opened up for the study of intra- and /or inter-site variability among animal bone assemblages.

ACKNOWLEDGEMENTS

We are very grateful to all those who have helped at various stages in this study. Particularly, we would thank Clive Orton and Paul Tyers of the Institute of Archaeology in London for their time, advice and patient computing help throughout the project. The London sites' data used in this project were collected under post-excavation programmes funded by English Heritage and made available by the Museum of London. The Stanwick (Durham) data were collected with funds granted from the North Yorkshire County Council and the Howard Trust.

BIBLIOGRAPHY

Armitage, P. L. (1982) Studies on the remains of domestic livestock from Roman, medieval and early modern London: objectives and methods. In: *Environmental Archaeology in the urban context*. (eds. Hall, A. and Kenward, H.), 94–106. CBA Research Report 43.

Bishop, Y.M.M., Fienberg, S.E. and Holland, P.W. (1975) *Discrete Multivariate Analysis*. MIT Press.

Gifford, D. (1981) Taphonomy and palaeoecology: a critical review of archaeology's sister disciplines. In: *Advances in archaeological method and theory*. (ed. Schiffer, M.), 365–438. Volume 4, Academic Press, New York.

Grayson, D.K. (1984) *Quantitative zooarchaeology*. Academic Press, London.

Greenacre, M.J. (1984) *Theory and Applications of Correspondence Analysis*. Academic Press, London.

King, A.C. (1978) A comparative study of bone assemblages from Roman sites in Britain. *Institute of Archaeology Bulletin*, 15, 207–232.

Maltby, J.M. (1984) Animal bones and the Romano-British economy. In: *Animals and Archaeology:4. Husbandry in Europe*. (eds. Grigson, C. and Clutton-Brock, J.), 125–138. Oxford BAR Int Ser 227.

Maltby. J.M. (1989) Urban rural variations in the butchering of cattle in Romano-British Hampshire. In: *Diet and Crafts in Towns*. (eds. Serjeantson, D. and Waldron, T.), 75–106. Oxford BAR 199.

Moreno Garcia, M. (1991) *A multivariate statistical analysis of archaeological bone assemblages in England*, unpublished M.Sc. thesis, University College London. Institute of Archaeology.

Moreno Garcia, M., Orton, C.R. and Rackham, D.J. (forthcoming) A new statistical tool for comparing animal bone assemblages.

Orton, C.R.and Tyers, P.A. (1990) Statistical analysis of ceramic assemblages. *Archaeologia e Calculatori*, 1, 81–110.

Orton, C.R. and Tyers, P.A. (1992) Counting broken objects: the statistics of ceramic assemblages. *Proc. British Academy*, 77, 163–184.

Orton, C.R. (forthcoming) Dem dry bones. In: *Interpreting Roman London: Essays in Memory of Hugh Chapman*.

Payne, S. (1972) Partial recovery and sample bias: the results of some sieving experiments. In: *Papers in economic prehistory*. (ed. Higgs, E.S.), 49–64. Cambridge University Press.

Rackham, D.J. (1986) Assessing the relative frequencies of species by the application of a stochastic model to a zooarchaeological database. In: *Database management and zooarchaeology*. (ed. van Wijngaarden-Bakker, L.), pp.185–192. Journal of the European Study Group of Physical, Chemical, Biological and Mathematical techniques applied to Archaeology, Research volume 40.

Ringrose, T.J. (1993) Bone Counts and Statistics: A Critique. *Journal of Archaeological Science*, 20, 121–157.

Schofield, J. (1987) *Museum of London: Department of Urban Archaeology Archive Catalogue*.

Serjeantson, D. (1989) Introduction. In: *Diet and Crafts in Towns*. (eds. Serjeantson, D. and Waldron, T.), 1–12. Oxford BAR 199.

Watson, J.P.N. (1979) The estimation of the relative frequency of mammalian species: Khirokhitia 1972. *Journal of Archaeological Science*, 6, 127–137.

APPENDIX 1: SITES REPRESENTED IN THE DATASET (OPEN CIRCLES ON THE PLOT).

CODE for sites on Figures 4, 5 and 6	FULL SITE CODE	SITE
BOP	BOP82	28–32 Bishopgate, EC2.
BSF	BSF86	Beddington Sewage Farm, Surrey.
CH	CH75	Chaucer House, Tabard St., SE1.
HOO	HOO88	Hooper St., EC1.
ILA	ILA79	Miles Lane, 123–7 Upper Thames St., EC4.
LOW	LOW88	85 London Wall, EC2.
LWA	LWA84	43 London Wall, EC2.
OPT	OPT81	2–3 Cross Key Court, Copthall Avenue, EC2.
RAG	RAG82	61–5 Crutched Friars, 1–12 Rangoon St., EC3.
SW	SW84II/SW85	Stanwick, County Durham.

APPENDIX 2: CODES OF TAXA PRESENTED IN THE TEXT (SOLID "CLOVER LEAF" ON THE PLOTS).

CODE	TAXON
EQU	Horse (*Equus caballus*)
BOS	Cattle (*Bos taurus*)
OVCA	Sheep/goat (*Ovis/Capra*)
OVI	Sheep (*Ovis aries*)
SUS	Pig (*Sus* dom.)
CSZ	Cattle size
SSZ	Sheep size
CANCAN	Dog (*Canis familiaris*)
FEL	Cat (*Felis catus*)

APPENDIX 3: CODES FOR BONE ELEMENTS PRESENTED IN THE DATASET (SOLID DIAMOND ON THE PLOTS).

CODE	BONE ELEMENT
SKL	Skull
MAN	Mandible
ATL	Atlas
AXI	Axis
CEV	Cervical vertebra
TRV	Thoracic vertebra
LMV	Lumbar vertebra
SAC	Sacrum
CDV	Caudal vertebra
SCP	Scapula
HUM	Humerus
RAD	Radius
ULN	Ulna
MTC	Metacarpus
PH1	First phalanx
PH2	Second phalanx
PH3	Third phalanx
INN	Innominate
FEM	Femur
TIB	Tibia
AST	Astragalus
CAL	Calcaneus
MTT	Metatarsus
LBF	Long bone fragment

13. Towards describing the nature of the chief taphonomic agent

Bob Wilson

1. INTRODUCTION

Environmental archaeology aims to reconstruct past human environments and seek explanations of ecological change. Yet frequently "man", as an ecological, cultural and taphonomic agent at ancient settlements, is left poorly defined and specified. While much factual information about our species is known when it comes to making a concise and informative, yet relatively complete, exposition on that nature of man and especially the causes of human action, the literature of environmental archaeology seems lacking in substance.

Of course, in archaeology the idea of investigating or defining human nature is more remote than for most natural and social scientists:

- Neither past human behaviour nor past mentalities can be observed directly
- Only fragmentary material remains of such activities remain in the ground

Investigating and defining human nature is largely done by reference to theories and paradigms of the sciences and humanities devoted to understanding present peoples and those of the limited historical past. As part social scientists we should, however, take an interest in such research. This paper outlines a general model by which major and characteristic outcomes of human organisation may be related to the elements of human motivation, the causes of observable behaviour.

2. DEVELOPING THE MODEL

In the general environmental and biological literature "man" tends to be defined, if at all, in biological and cultural terms (compare Watson and Watson, 1969: 17–21; Simmons, 1974: 3–48, Goudie, 1989: 337–40 and Mielke, 1989: 241–2) but less well in psychological and social terms. There appears only one sizeable precedent in the archaeological literature and that is Graeme Clark's book on the identity of man (1983). I disagree with the orientation of his later discussion but in one diagram, however, he lists and links human phenomena which make the complexity of man, culture and ecology more explicit – a useful beginning to deeper analysis (Clark, 1983: Fig. 17). Other works helpful in the approach to this are those of Johnson (1983) in the relevant field of geography and Lloyd (1986, 1993) in the historical field and the modelling of human nature.

Study of human nature appears concerned with two problematic characterisations: 1) the constitution of it and 2) the way in which it operates, i.e. the study of motivation. Understanding human motivation seems to require a consideration of the interplay between relatively unchanging structure on the one hand and processes of changing and varying structure on the other. Motivation is correlated with the changes from existential structure to existential structure. It is very much based upon the needs, wants and goals of people which maintain human structures and processes of existence. Figure 13: 1 and subsequent text summarises the important steps in the characterisation of human nature.

To begin with it is not important to list and discuss the separate needs and wants of people which vary greatly with culture. Instead it is instructive to deal:

1) with the components by which needs, wants and goals are obtained and
2) with the phenomenon that, when needs and wants are achieved, satisfaction and often a variety of positive emotions are felt or experienced.

This outcome can be described as 'being in the world' (Figure 13: 2–3). Perception, logic, values, ideals and overt behaviour are involved.

Also, in reality, at least in modern times, there are many occasions when need and want satisfaction are

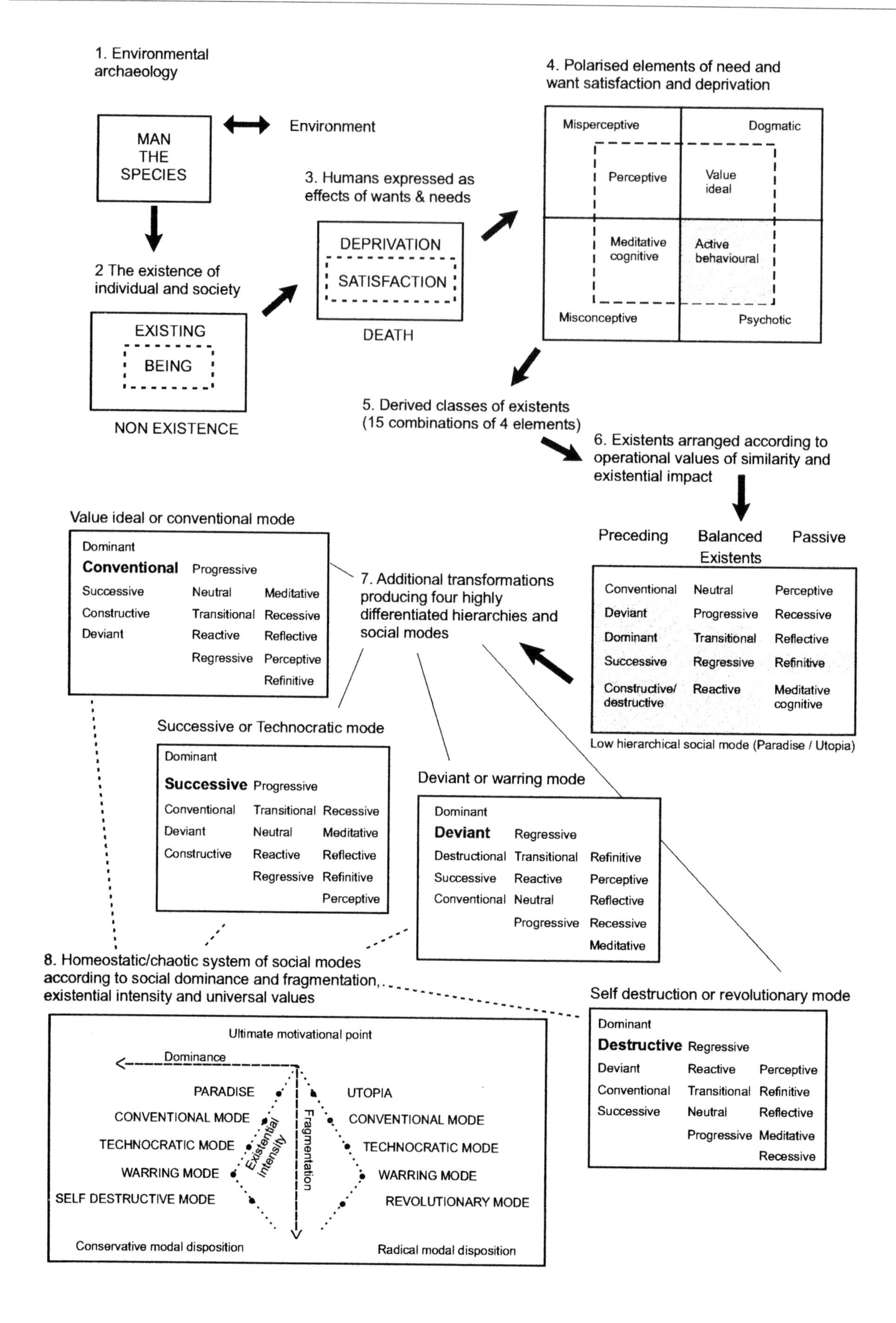

Figure 13.1: A model of human characteristics and behaviour.

partially or wholly *not* obtained. This need deprivation and existential intensity are associated with other sets of experiences, feelings and, often, negative emotions ('merely existing in the world'). So the overall process of human life is better described, although not defined completely, by four moderately complicated components or *elements* that describe simple need and want satisfaction:

1) Perceptive – misperceptive element
2) Active or appropriate behavioural– psychotic element
3) Value ideal – dogmatic element
4) Meditative cognitive – misconceptive element

The positive poles of the elements contribute to need satisfaction and the negative poles are associated with deprivation and existential intensity. Another representation of these elements is given in Figure 13: 4. Note that three elements are essentially internal and mental activities while the fourth, the active behavioural, tends to be external and observable. Much therefore of these inferred elements is difficult to verify.

The four elements are always altering processes, albeit retaining their defining characteristics, in every individual, especially as needs renew themselves. They vary greatly from person to person. Nonetheless, because they have their own characteristics, they can be considered as building blocks of higher 'structures' of existence and being which relate individual, often hidden, and social, often observable, outcomes. Here 'structures' could refer in part to the external material structures of culture such as buildings or utensils (remains of which are observable) since they are produced from functional, symbolic and value ideal rationalisations. Mostly, however, structures appear an amalgam of mental and physiological mechanisms comprising the causality which makes up, and relates, higher and lower levels of human organisation. Certainly the structures inferred by me relate to abstract means by which social processes, common to more than one culture, can be understood.

Partly at the level of the individual, but more commonly at the level of social action (an improvement from 25% at the elemental level of understanding to a theoretical 53% of observable human phenomena), there is higher human organisation – the units of which are called 'existents' (Figure 13: 5) As individuals and social groups set about need and want satisfaction the four elements of need satisfaction occur in up to 15 different combinations in activities, events or situations and each actual combination or 'existent' present has an overall characteristic nature.

Class I

– substantial presence of any one element above, and limited presence of, the others.

There are four existents in this class and these are almost identical in their character as elements except that there is a change of explanatory emphasis in their names. For instance, the value – ideal element becomes known as the 'conventional existent' in order to emphasise the employment of values and norms according to the consensus of a ruling or democratic group in a culture. A second example is the active behavioural–psychotic element which becomes known as the 'constructive– destructive existent' since its effect is to assist or deny need satisfaction and structural maintenance. The other two existents are referred to simply as the 'perceptive' and 'meditative' existents.

Class II

– presence and predominance of any two elements.
Six existents are possible and their characters are determined from the combination of elements present. They are called 'neutral', 'reactive', 'dominant', 'reflective', 'progressive' and 'regressive' existents. As an important explanatory example the 'dominant existent' consists of the substantial presence of the value ideal and the active behavioural elements. Leadership, political qualities and similar determining roles are characteristic of this existent since behaviour based upon conventional values and ideals is decisive and efficient compared to outcomes of activity associated with other existents. At existential extremes, however, the dominant existent is expressed as domineering action and dictatorial leadership.

Class III

– predominance of three elements
Four existents are possible – 'successive', 'recessive', 'deviant' and 'refinitive'. The 'successive existent' is of interest to archaeologists since it includes technological, exchange and commercial activities based upon the thoughtful application of values and the evaluation of ideas. The 'deviant existent' is quite distinctive too. It concerns activities in values, perceptions and misperceptions and in behaviour which are different from the 'conventional existent' and which, at existential extremes, constitute 'immoral' or 'criminal' actions.

Class IV

– presence of all four elements
One existent only – the 'transitional' existent, 'transitional' since need satisfaction of biological and cultural activity is more likely than with other existents where one to three elements are absent. However, at extremes of misperception, misvaluation, illogical thinking and inappropriate behaviour, needs and wants will not be satisfied. Apathy and death may result under such circumstances and is true of all existents at existential extremes.

Since the fifteen existents are argued to exist in any culture or society, they may possess certain relationships with each other, particularly where the overall social context (*mode*) and the hierarchical order of existential impact and social influences of existents are considered. Five main existential hierarchies and associated social modes are possible (Figure 13:6 and 7), according to fairly basic operational values assigned to the elements present in each existent. In particular, value – ideal and active behavioural elements are postulated to have greater existential impact than the presence of perceptive and meditative elements. This leads to a first, and little or moderately differentiated, existential hierarchy structured as follows:

'preceding existents'	*'balanced existents'*	*'passive existents'*
Conventional	Neutral	Perceptive
Deviant	Progressive	Recessive
DOMINANT	Transitional	Reflective
Successive	Regressive	Refinitive
Constructive-destructive	Reactive	Meditative

Only the dominant existent takes precedence among the other 'preceding' existents which share equal value and influence, particularly on the dominant (or leadership) existent. This hierarchical arrangement is regarded, at least in theory, to correspond to paradise or utopian society or mode of society. Possibly this is a metaphysical category although certain traditional African and other cultures appear to have had low hierarchical structures resembling aspects of the above existential structure, e.g. Category I cultures (Kent, 1990: 130–32 and fig. 9.2; Turnbull, 1978).

With further operational value differentiation between existents, four more highly differentiated existential hierarchies and their social modes are probable outcomes of human activity. These are :

1) Conventional or Value Ideal mode
2) Successive or Technocratic mode
3) Deviant or Warring mode
4) Self destructive or Involutional/ Revolutionary mode

They are named after, and substantially determined by, the existent which is closest to the dominant or leadership existent. Thus the order and proximity of existents is an approximate guide to their influence on decision making and behavioural outcome of each society or mode. From comparison of the expressive names of existents and their varied rankings in the lists of Figure 13:7, a modest idea of the general nature of each mode of society may be obtained.

An example of the fairly broad interpretation of one of these hierarchies is:

In this mode of existential society the Deviant ('criminal and immoral') existent is the most influential on the leadership and war against other cultures is an inevitable outcome; the most extreme of deviant human behaviour. Destructive (negative pole of constructive – destructive existent) and Regressive ('backward looking and acting') existents are next most influential. Thus there is an emergence and strong influence of (by universal standards) criminal and immoral behavioural phenomena although the majority of the population may not view their actions by these standards. Thus 'normal' values and behaviour become variously distorted, prejudiced and inconsistent with conventional (relatively universal) values. Aggressive attitudes and normally psychotic behaviour become encouraged and channelled as effectively as possible in warfare against the enemy. In modern society at least, censorship, secrecy, self justifying religion, propaganda, jingoism, arms manufacture, black-marketing, looting and profiteering are liable to upset conventional social organisation as the deviant mode develops in war and probably becomes increasingly corrupt.

Leadership tends toward dictatorship and away from democracy. Social justice is more limited and punitive than in conventional and successive social modes – witness the treatment of pacifists and spies. Ecological restraint is forgotten.

Conventional social needs and well-being are thwarted and deprived by costs of war, giving poor working, 'living' and fighting conditions. Existential feeling and deprivation are intensified throughout the whole population. Although the times of active fighting may generate high excitement, courageous resolve and delusion of 'being' in the flux of battle, these are offset by long periods of boredom, a sense of futility, grief and other misery from wounds, madness and death of others" (Wilson, unpublished a (slightly modified)).

Of course such description and interpretation are mainly common sense but are justified as part of the attempt to categorise, objectively, human nature as an intelligible system rather than implicating war as a largely irrational and aberrant outcome isolated from social and cultural causes in more desirable modes. Secondly, the above description is justified here since much of the psychological phenomena described above underlay the extremes of aggressive and exploitative behaviour involved in past territorial invasions, livestock raiding and ideological conflicts.

The modes of society and their inter-changeability can be listed in order of their increasing existential intensity, from modes of well-being and need satisfaction to states of war and revolution where need deprivation, apathy and death are common. This relationship can be shown as:

Ultimate motivation point	
[.]	
Paradise.	.Utopia
Conventional mode.	.Conventional mode
Technocratic mode.	.Technocratic mode
Warring mode.	.Warring mode
Self destructive mode.	.Revolutionary mode
Conservative modal disposition	*Radical mode disposition*

An elaborated scheme is shown in Figure 13:8. Within this scheme any culture may not exist for long in any one mode or type. It is postulated to fluctuate from mode to mode according to a complexity of cultural factors. It can move vertically and is paralleled by correlated existential intensity but with basically the same set of conventional values, albeit beneficial or corrupt. It also can move laterally at high or low levels from relatively conservative conventional values to a significantly different set of conventional values in a more radical modal disposition of social modes (which become conservative over time before further radical change).

Potentially the scheme represents a homeostatic system in which a culture moves existentially according to the self awareness of its people to their cultural values in relation to universal social values. At present most cultures appear to move chaotically from mode to mode rather than succeeding for long at peaceful resolution and self control leading to states of lesser existential intensity. Thus each culture will take its own largely unique existential pathway, the details of which are related to the perceptiveness–misperceptiveness,valuation–misvaluation, logics valid or false, and appropriate–inappropriateness of behaviour, in phenomena associated with its particular sets of values as well as the strength of its cultural traditions. Plainly these processual structures are adaptively related in the knowledge of the particular environment each culture lives in, creating its own ecology, although typically ceremonial and other ritual processes will vary any simple relationships with natural systems.

3. CONCLUSIONS

This model or paradigm of human nature is, of course, largely based on the psychological and sociological conceptions of modern societies (eg. Douglas, 1964; Giddens, 1993) rather than on historical and archaeological knowledge. Nonetheless the elements, existents and modes underlying this general paradigm are postulated not to have changed during evolution of 'modern' man (otherwise the definition of the species would have to change). However, the duration of cultures in each mode will have varied. The impression is that many early cultures and some modern ones existed or exist mainly in the Conventional mode while later, more complex 'civilisations' and modern Western society tended or tend to exist in Technocratic and Warring modes. Thus the quality and detail of human existence probably changed during human evolution.

The intent of this paper has not been to present either a definition of man the species, or a complete model/paradigm of human nature, but to illustrate briefly important parts of one and indicate their often invisible causal linkages in order to open up academic discussions on the validity of this type of approach. Even with this fairly sparse summary is contributed additional complexity to archaeological interpretation which may prove useful in considering the social background in which the usual forces and factors of taphonomy are discussed.

For example, scavenging, rubbish clearance and butchery, which appear to be the three major taphonomic processes causing the spatial patterning of animal bones, have been discussed at some length for the Iron Age settlement at Mingies Ditch, Oxfordshire (Wilson, 1993). With the limited site evidence, there is little elaboration in the report of other cultural factors in bone deposition, apart from some discussion of kill-off patterns, livestock management and exchange of animals. Clearly, however, the artefactual and ecofactual patterning at Mingies Ditch scarcely reflect the totality and variety of social and cultural events during the life of the settlement and which may well have involved existential extremes of livestock raiding, blood feuds and inter-tribal warfare besides the normally inferred subsistence activities (Ross, 1970; Cunliffe, 1991; Lucas, 1989).

However, with an adequate and explicit published paradigm of human nature, a substantial understanding of the diversity and complexity of existential factors and forces is automatically implicit and justifiable in any truly prehistoric site interpretation although the causal contributions of most elements, existents and modes remain largely invisible in the archaeological record.

BIBLIOGRAPHY

Clark, G. (1983) *The identity of Man*. London.
Cunliffe, B. (1991) *Iron Age communities in Britain*. London.
Douglas, M (ed) (1964) *Man in society*. London.
Giddens, A. 1993 (1989) *Sociology*. Oxford.
Goudie, A. 1989 (1984) *The nature of the environment*. Oxford.
Johnson, R.J. (1983) *Philosophy and human geography*. London.
Kent, S. (1990) A cross-cultural study of segmentation and the use of space. In: Kent, S. (ed.) *Domestic architecture and the use of space*. Cambridge.
Lloyd, C. (1986) *Explanation in social history*. Oxford.
Lloyd, C. (1993) *The structures of history*. Oxford.
Lucas, A.T. (1989) *Cattle in ancient Ireland*. Kilkenny.
Mielke, H.W. (1989) *Patterns of life*. London.
Ross, A. (1970) *Everyday life of the pagan celts*. London.
Simmons, I.G. (1974) *The ecology of natural resources*. London.
Turnbull, C.M. 1978 (1976) *Man in Africa*. London.
Watson, R.A. and Watson, P.J. (1969) *Man and nature*. New York.
Wilson, R. (1993) Bone reports. In. T.G. Allen and M.A. Robinson. *The prehistoric landscape and Iron Age enclosed settlement at Mingies Ditch, Hardwick-with-Yelford, Oxfordshire*. Thames Valley Landscapes: the Windrush Valley, 2, 123–34 and 168–204.
Wilson, R. unpublished a Human identity: a rainbow of meaning.
Young, J.Z. (1971) *An introduction to the study of man*. Oxford.

Papers presented at the conference but not submitted for publication

Bone weight loss, taphonomy and the tale of a fish midden

James H. Barrett, then of: Department of Archaeology, The University of Glasgow.

This paper explores the effect of taphonomic processes on the weight of archaeological recovered fish bones. Linear measurements of bones from Robert's Haven, a medieval fish midden near John O'Groats, Scotland, are used to predict what each bone might weigh if fresh, dry and complete. These predictions are then compared to the actual dry weight of the specimens. The results help elucidate the impact of taphonomic processes on the Weight Method, and provide an avenue for comparing the degree of bone destruction in different fish bone assemblages.

Taphonomy and the palynology of cave sediments

Geraint Coles, Department of Archaeology, Univeristy of Edinburgh.

Palynological analysis of cave sediments has made significant contributions to understanding environmental change in limestone areas otherwise devoid of suitable sites for palaeoecological investigation (eg. Leroi-Gourhan, 1965; Peterson, 1976; van Zinderen Bakker, 1982; Hunt and Gale, 1985; Coles, 1987). However, many problems exist in cave palynologoy (cf. Turner, 1985); poor understanding of the taphonomic pathways and processes governing the formation of pollen and spore assemblages being the most prominent (cf. Coles et al., 1989). This paper sets out to review our understanding of the mechanisms by which pollen and spores are recruited to the interior of caves and the post depositional processes which govern their preservation. Examples of particular processes are given.

Taphonomic aspects of zooarchaeology on Bahrain

Keith Dobney and Brian Irving, Environmental Archaeology Unit, York and Human Environment Group, Insititue of Archaeology, London.

Excavations at the middle Dilmoun site of Saar, on Bahrain, have produced a large and diverse zooarchaeological assemblage. An extensive and systematic quantitative wet and dry-sieving programme, and the presence of on-site specialists during excavation, has meant that a number of important taphonomic questions could be addressed.

All the standing buildings are filled with sand which makes archaeological interpretation of their varying stratigraphic relationships particularly difficult. This overlies what are considered to be primary occupation deposits within the various buildings and are thought to represent abandonment episodes when the buildings were used as convenient dumping areas for the town's rubbish.

Preliminary analysis of the proportions of fish, mammal and bird remains, coupled with varying bone preservation, has shown that much variation exists between these archaeologically very similar deposits in different parts of the town.

Telling it like it wasn't: Living conditions in Viking-Age Dublin

Siobhán Geraghty, then of: National Museum of Ireland, Kildare Street, Dublin 2, Eire

Early medieval houses and towns were different from those of today, and their archaeological remains indicate a very different flora and fauna. These differences are often used to portray a very graphic picture of a dirty, smelly, medieval lifestyle which has become firmly fixed in the mind of the general public.

The results of close examination of the floor deposits

from Viking-Age Fishamble Street suggests that occupants made conscious efforts to maintain a standard of order and cleanliness within their houses, and that the apparent evidence to the contrary arises from a combination of the materials used, and the vagaries of human behaviour.

A study of late 20th century habits and living conditions indicates that despite differences in detail, much remains the same, and that the popular image of the middle ages may be a distortion.

The Dexter – a rôle model for archaeological cattle ?

L.J. Gidney, Department of Archaeology, University of Durham

The Dexter is the smallest surviving native British breed of cattle. Bones from this breed are of comparable size to archaeological specimens from later Prehistoric and Romano-British sites. However, the Dexter is often discounted as a reference animal for archaeological collections because of the dwarf gene within the breed. I am collecting skeletons of modern Dexters, most of which are not dwarfs, to ascertain how closely the breed does resemble the archaeological stock. This paper will also briefly examine the productivity of a small commercial dairy herd kept in the traditional manner in the Pennines. This will serve as an example of the potential output of the archaeological animals.

Groundwater geochemical modelling in archaeology – the need and the potential

A.M. Pollard, Department of Archaeological Sciences, University of Bradford

Groundwater modelling – the use of computer models to predict the interaction between inorganic (and organic) geochemistry is used to study problems such as the long-term stability of buried vitrified waste, or the potential toxicity of pollutants released into the natural environment. So far, it has only been used once in archaeology, to predict the corrosion behaviour of copper alloys in the burial environment (Thomas, 1990). The situations where modelling could help to transform an archaeological study from observational to predictive are endless – not only modelling metal-groundwater interactions, but also ceramics and glasses (and even tephra) in natural groundwaters. One of the most interesting applications is to model the diagenesis of bone, but this is also more difficult because of the large initial organic component.

Unfortunately, although modelling programs are widely available, the applications are not that simple. The programs require detailed data on factors such as the thermodynamic properties of all the species involved, and a knowledge of the redox and pH of the appropriate environment. In reality, many approximations have to be made, and it is always essential to check the models against laboratory experiments and, ultimately, against real-life measurements. This paper will review briefly the potential of groundwater modelling applications in archaeology.

Post-depositional modification of cave pollen spectra

C.O. Hunt, Department of Geographical and Environmental Sciences, University of Huddersfield.

Six major and three minor temperate episodes can be recognised palynologically in the infill sediments of Robin Hood's Cave, Creswell Crags, Derbyshire, but the sequence of pollen zones associated with each episode is unlike those registered at open-air sites of comparable age. Reasons for the discrepancy can be partially attributed to biogeography and depositional (taphonomic) factors. Also very important are post-depositional processes of bioturbation and selective preservation. The impact of these processes on pollen assemblages are explored in this contribution.

Taphonomy of molluscs in caves: a conceptual model

C.O. Hunt, Department of Geographical and Environmental Sciences, University of Huddersfield.

Comparatively few mollusc assemblages have been reported from archaeological caves, but it is already clear that taphonomic factors have a significant impact upon assemblage composition. In this contribution, the taphonomic factors underlying mollusc assemblages in caves are explored.

Abstracts accepted but participants unable to attend

Taphonomic analysis of the fauna from the Middle Palaeolithic Open Site of Wallertheim (Germany)

Sabine Gaudzinski, Forschunsstelle Altsteinzeit, Schloss Monrepos, 56567 Neuwied, Germany

The open site of Wallertheim was excavated in 1927 and 1928. It has since become a relatively well known site, due to the rich faunal remains and the stone artefacts. Results of recent geological, malacological and palynological analyses as well as studies of the mammalian fauna place the main find-horizon at the end of the Eemian Interglacial period or in an early Weichselian Interstadial phase, i.e. roughly to an early phase of oxygen isotope stage 5.

The mammalian fauna of the main find level consists of nine species. *Bison priscus*, the steppe bison, dominates the faunal assemblage with an MNI of 52 individuals. Two species of *Equus* (*E. germanicus* and *E. przewalskii*) were also present and indicate 13 individuals altogether.

The Wallertheim site represents part of a riverine environment, where hominid and non-hominid agents played a role in the formation of an assemblage of large mammals.

A comparative analysis of the bones of the steppe bison and horse, the two species dominating the faunal assemblage, was undertaken. Features such as certain bone surface modifications, the skeletal element representation, the degree of bone fragmentation and the age profiles were examined.

As a result of the taphonomic analysis it is argued that only one species of the death assemblage, *Bison priscus*, can be convincingly related to human activities. According to this interpretation all the other species have to be considered as part of the natural background fauna. Results like these qualify the implications of taphonomic analysis for the interpretation of Middle Palaeolithic subsistence strategies.

The significance of body part representation in hyaena den assemblages

Liora Kolska Horwitz, Israel Antiquities Authority, POBox 586, Jerusalem 91004, Israel

Scavenged bone assemblages were recovered from a number of sub-recent hyaena dens in Israel and examined to determine bodypart representation after Stiner (1992). Significant inter-species differences were found in the ratio of cranial to limb bones. The ratio of cranial to post-cranial bones in small animals (sheep, goat and dogs) was 5 times higher than in large animals (equids and camels). This may reflect selection at the scavenge site, with striped hyaenas intentionally removing crania or else having only crania available after other carnivores had abandoned the carcass. Alternatively, it may reflect hyaena activity within the dens with almost complete destruction of the post-cranial skeleton of small animals but some preservation of the cranial bones.

Reference

Stiner, M.C. (1991) Food procurement and transport by human and non-human predators. *Journal of Archaeological Science*, 18, 455–482.

Posters

Environment and archaeology in Scotland: current work by AOC (Scotland) Ltd.

AOC (Scotland) Ltd., The Schoolhouse, 4 Lochend Road, Leith, Edinburgh, EH6 8BR

This poster introduces some current archaeological projects being undertaken by AOC (Scotland) Ltd. and mainly funded by Historic Scotland. In each project post-excavation analyses form a key part of an integrated project design and are used primarily to test field interpretations. An understanding of site formation processes is recognised as central in making informed choices about resource allocation and about the level of interpretation which can be applied to any set of results.

In the Lairg project, Sutherland, the emphasis in the project design is on the archaeology of a changing landscape, rather than the more traditional monument based approach. Investigations examine how successive farming communities have shaped the land and have been shaped by it.

Excavation of a Dark Age crannog at Buiston, Ayrshire, has given an opportunity to study the occupation history, living conditions and economy of this remarkable site. This study has been greatly aided by the waterlogged conditions which have preserved a wealth of organic remains not frequently encountered in Scottish rural sites. Good chronological resolution is possible on this site, enhancing the potential for detailed analysis and interpretation; dendrochronological studies have demonstrated a complex history of construction and repair.

At St. Boniface, Orkney, coastal erosion has prompted investigation of a huge medieval farm mound. Site formation processes are key to understanding this site, which has been greatly modified over time, and micromorphological study of sediments has proved particularly fruitful.

Sheep: teeth and death

L. J. Gidney, Department of Archaeology, University of Durham,

The mortality pattern of the Zanfara flock of Manx Loghtan sheep (established 1985) will be examined using Mandibular Wear Scores. Patterns for male and female deaths, both deliberate and natural, will be established for animals of known sex and age at death.

Indication of man-induced soil erosion in sediment cores from Lake Constance (SW Germany) since c.5000 BC

Wolfgang Ostendorp, Limnologisches Institut, Universität Konstanz, D-7750 Konstanz, Germany

Three sediment cores from Lake Constance-Untersee were analysed for magnetic susceptibility, carbonates, organic carbon, total phosphorus, quartz, Mg and Sr. One of them, situated near to the mouth of an inlet shows signals that can be correlated with the onset of settlement phases in the hinterland, beginning with the Bandkeramik settlement at c.5500 to 5000 BC. A clear-cut connection with climatic changes during the Holocene could not be stated.

Sea level changes, coastal morphology and sediment taphonomy in northwest Scotland: implications for interpreting cultural activity from the palaeoecological record.

Ian Shennan, Jim Innes, Antony Long and Yong Quiang Zong, Department of Geography, University of Durham

The coastal zone of northwest Scotland contains a considerable archaeological record of prehistoric human settlement and activity. Sea-level fluctuations and crustal movements during the prehistoric period have, however, combined to produce a complex history of coastal changes in this area, which will have influenced prehistoric land use and archaeological site location. Adjacent coastal locations will have experienced radically different sediment regimes ranging from fully marine to fully terrestrial, governed by each site's morphology and relation to contemporaneous sea level. This is illustrated with stratigraphic and palaeoecological data from a series of isolation basins near Arisaig. Positive and negative relative sea-level movement would have caused major changes in the dominant environmental conditions and in sediment source areas at individual sites at different times. This, together with hydroseral succession within site catchments, will have significantly altered the taphonomy of microfossil assemblages contained within sediments at coastal sites. Interpretation of these assemblages in terms of prehistoric human activity therefore requires a sound knowledge of sedimentary palaeoenvironments as well as catchment vegetation history.